Junkyard Planet

Junkyard Planet

Travels in the Billion-Dollar Trash Trade

ADAM MINTER

BLOOMSBURY PUBLISHING

NEW YORK · LONDON · OXFORD · NEW DELHI · SYDNEY

BLOOMSBURY PUBLISHING
Bloomsbury Publishing Inc.
1385 Broadway, New York, NY 10018, USA

First published in the United States 2013
This paperback edition published 2015

Portions of the preface and chapter 13 were originally published in different form
in the *Atlantic* as "The Chinese Town That Turns Your Old Christmas Tree
Lights into Slippers" and "Scrapped."

Portions of chapters 5, 9, and 10 were originally published in different form
in *Scrap* as "The Future of Brass City," "China's Plastics Frontier," and
"The Industrial Revolution."

A section of chapter 14 was originally published in different form in *Recycling
International* as "A Happy Marriage of Processor and Machine Maker."

ISBN: HB: 978-1-60819-791-0; PB: 978-1-60819-793-4; eBook: 978-1-60819-792-7

Library of Congress Cataloging-in-Publication Data

Minter, Adam, 1970–
Junkyard planet : travels in the billion-dollar trash trade /
Adam Minter. — First U.S. edition.
pages cm
Includes index.
ISBN 978-1-60819-791-0 (hardcover)
1. Refuse disposal industry. 2. Refuse and refuse disposal. 3. Scrap materials.
4. Recycling (Waste, etc.) I. Title.
HD9975.A2M495 2013
338.4'7363728—dc23
2013011750

6 8 10 9 7

Typeset by Westchester Book Group
Printed and bound in the U.S.A. by Sheridan, Chelsea, Michigan

To find out more about our authors and books visit www.bloomsbury.com and
sign up for our newsletters.

For my grandmother, Betty Zeman.
She scrapped.

Rusted brandy in a diamond glass
Everything is made from dreams . . .

—TOM WAITS, "TEMPTATION"

Contents

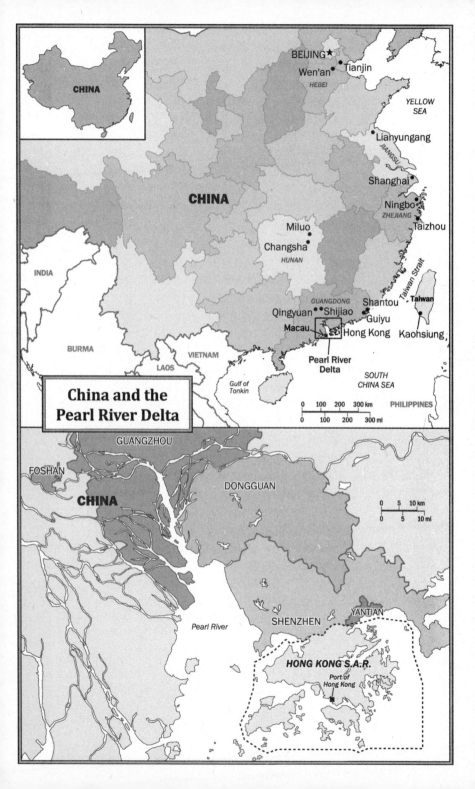

China and the Pearl River Delta

A Note on Numbers

It's impossible to write a book about an industry without statistics. So this book includes statistics about the scrap industry. Some of those numbers were tabulated by governments, some by businesses, and some by individuals. Nonetheless, all but a few of these statistics should—at best—be considered estimates.

As I'll demonstrate through this book, the global recycling and refurbishment industry is a notoriously difficult business to turn into clean data. In part, this is due to the nature of the product. Scrap is often a byproduct of something—garbage—that's rarely counted in a precise manner. And in developing countries, it's rarely counted at all.

In cases where scrap can be counted—say, out of a shipping container—there are additional problems. For example, shipping containers filled with scrap metal often contain mixtures of many different metals, some of which have never been precisely tabulated. In addition, those mixed metal loads are often declared falsely to governments, for the purpose of tax evasion. The practice is widespread enough to seriously impact the reliability of any and all scrap import and export statistics.

Finally, readers will note that I rarely cite numbers tabulated by environmental organizations, despite their widespread use and circulation in media. The reason is simple: scrap recycling is a business, and the best data—insofar as it exists—comes from organizations actually involved in the collection, shipping, and disposition of recyclables.

Introduction

A single strand of burned-out Christmas tree lights weighs almost nothing in the hand. But a hay-bale-sized block? That weighs around 2,200 pounds, according to Raymond Li, the fresh-faced but steely general manager of Yong Chang Processing, a scrap-metal processor in the southern Chinese town of Shijiao.

He would know.

I am standing between him and three such bales, or 6,600 pounds of Christmas tree lights that Americans tossed into recycling bins, or dropped off at the Salvation Army, or sold to someone in a "We Buy Junk" truck. Eventually they found their way to a scrapyard that pressed them into a cube and shipped them off to Raymond Li's Christmas tree light recycling factory.

Raymond is anxious to show me how it works.

But first off, he needs to tell me that, though 6,600 pounds might seem like a large volume of American Christmas tree lights to find in a small Chinese village, it isn't. Mid-November is actually low season for buying imported old Christmas tree lights. High-season starts after the New Year and reaches its peak in the spring, when Americans in the northern states start to empty their homes and garages of the pesky tangles. Those who take them to the local recycling center or sell them to the local scrapyard most likely have no idea where they're going next. But I do: right here, to Shijiao, China, population maybe 20,000. Raymond Li tells me that his company recycles around 2.2 million pounds of imported Christmas tree lights per year, and he estimates that Shijiao is home to at least nine other factories that import and process similar volumes. That's 20 million pounds annually, conservatively estimated.

How did an anonymous village in southern China become the Christmas Tree Light Recycling Capital of the World? Here's one answer: Shijiao is within driving distance of thousands of factories that need copper to make things like wires, power cords, and smartphones. Those factories have a choice: they can use copper mined in far-off, environmentally-sensitive places like the Brazilian Amazon. Or, alternatively, they can use copper mined from imported Christmas tree lights in Shijiao.

But Raymond's answer as to how Shijiao achieved its odd status is much simpler: "People wanted to make money," he says softly, his distant gaze pointed away from me. "That's all."

Raymond knows the history as well as anyone, and he tells it quickly, with no adornment. In the early 1990s economic opportunities were limited in Shijiao: you either farmed, or you left. The area lacked decent roads, an educated workforce, or raw materials. All it really had was space—vast, remote space. And as it happens, remote space, a box of matches, and some fuel are all you need to extract copper from a pile of old Christmas tree lights. Just douse the wire, set it on fire, and try not to breathe the fumes as the insulation burns off.

Raymond leads me into a cramped office where cloudy windows face Yong Chang Processing's factory floor. I'm offered a seat on a dusty leather sofa. Taking the seat to my right is Cousin Yao, brother to Raymond's wife, Yao Yei, who is seated across from me. Low-key Raymond, native of Shijiao, takes a seat beside his wife. It's a family business, they tell me, and everyone helps out.

I glance out the window at the factory floor, but from the sofa's low vantage point I can't see past additional piles of scrap wire (not Christmas tree lights) worth tens of thousands of dollars that Raymond imported a few days ago. If Raymond feels like it, he's flush enough to buy millions of dollars' worth of U.S. scrap metal per month. That may seem like a large number. But really, it's not. The global recycling industry turns over as much as $500 billion annually—roughly equal to the GDP of Norway— and employs more people than any other industry on the planet except agriculture. Raymond Li is big in Shijiao, but here in Guangdong Province, the de facto headquarters of China's recycling industry, he has many peers.

We chat more about the history of Shijiao, its wire recyclers, and how it's changed the lives of thousands of former farmers. Then, abruptly, Cousin Yao announces that he received a degree in engineering from a top university. Rather than join a traditional manufacturer, he says, he returned to Shijiao to join Raymond's scrap business. He could have gone anywhere, he could have done other things. China, after all, doesn't lack opportunities for engineers. But Cousin Yao knows a better opportunity when he sees one, and scrap metal was that opportunity. As he and Raymond see it, China's economy is expanding quickly, and its government planners and businessmen are desperate to find copper, steel, paper pulp, and other raw materials to feed the factories that drive the growth. Copper mines are great, but Raymond and his family don't have the money or connections to open a copper mine. Then again, why would they want to do that, when there's an endless supply of perfectly recyclable and reusable copper—worth billions!—available in the junkyards and recycling bins of America?

Raymond lights a cigarette and explains that he didn't have Cousin Yao's choices. Fifteen years ago he was twenty-seven and working as a laborer in a dead-end job at a paint and chemical factory. "I wanted to be rich and successful," he explains softly. "So I joined the scrap business." His wife's family was already engaged in scrapping on a small scale. They knew how and where to get recyclable scrap, and better yet, they knew the potential that foreign throwaways have to make a family rich—much richer than rice farmers, storekeepers, and office workers.

Since Raymond's fateful decision, China's raw material needs have only grown, and so has Raymond's business. Take, for example, China's demand for oil. As late as 2009 visitors to Shijiao were confronted with clouds of black smoke churning off giant piles of burning wire (not just Christmas tree wire, either). The rubber insulation was worthless; back then it was the copper that everyone wanted, and burning was the quickest way to liberate it. Then something important happened. Chinese started buying cars, driving up the price of oil and things made from oil—like the plastic used to insulate Christmas tree lights. As the price of plastic rose, Chinese manufacturers started looking for alternatives to "virgin" plastic made from oil. The most obvious solution was the cheapest: instead of burning plastic off copper wire, figure out a way to strip and recover it for reuse. Wire insulation isn't the highest quality plastic, but it's good enough to make simple products like . . . slipper soles! These days, the biggest customers for Raymond's Christmas tree insulation are slipper sole manufacturers.

Of course, getting from Christmas lights to slipper soles isn't easy or obvious. It took Cousin Yao more than a year of tinkering and testing to get Yong Chang's Christmas tree light recycling system right. I look around the room and ask whether I might see it. Raymond nods, and we walk out to the factory floor.

The process begins with workers paid as much as $500 per month to toss handfuls of Christmas tree lights into small shredders (they look like wood chippers). With thunderous groans, the shredders pulverize the tangles into millimeter-sized bits of plastic and metal and then spit them out as a mudlike goop. Next to those shredders are three vibrating ten-foot-long tables. As workers shovel the goopy shredded lights onto their surface, a thin film of water washes over them, bleeding out very distinct green and gold streaks. I step closer: the green streak is plastic, and it washes off the table's edge; the gold streak is copper, and it slowly moves down the length of the table until it falls off the end, into a basket, 95 percent pure and ready for remelting.

The principle at work is simple: think of a streambed covered in gravel. A flowing current will pick up the smaller pieces and carry them downstream quickly, while the bigger piece, the rocks, will stay in place, only occasionally moving. The same physics is at work on Raymond's tables, only it's not gravel that's carried away, it's Christmas tree light insulation.

Recycling. A generation of Americans defines it as: the act of sorting cans from bottles from cardboard from newspapers and setting them out on the curbside, or down in the trash room, for somebody to pick up. It's an act of faith, a bet that the local recycling company or trash collector is as committed to doing the environmentally sound thing as the person who sorted the recyclables in the first place. But what is that right thing? And is it really recycling if your carefully sorted newspapers, cans, and bottles are shipped off to Asia?

Definitions are important, and from the standpoint of the recycling industry, what most Americans think of as "recycling" is actually more akin to harvesting. That is, a home recycler harvests cardboard from trash and other recyclables, and a paper mill recycles that used cardboard into new cardboard. Recycling is what happens *after* the recycling bin leaves your curb. Home recycling—what you most likely do—is just the first step. Nonetheless, it's the key step: no machine can harvest recyclables from your trash as cheaply and efficiently as you can. In fact, compared to harvesting, the actual recycling is often the easy part. After all, the process by which old paper is transformed into new paper is centuries old; turning old computers into new ones is more difficult, but only because the machines are complicated to pull apart. But harvesting enough paper to make a paper mill run? That's difficult. Finding enough computers to justify opening a computer reuse or recycling business? That might even be harder.

This book aims to explain why the hidden world of globalized recycling and reclamation is the most logical (and greenest) endpoint in a long chain that begins with the harvest in your home recycling bin, or down at the local junkyard. There are few moral certainties here, but there is a guarantee: if what you toss into your recycling bin can be used in some way, the international scrap recycling business will manage to deliver it to the person or company who can do so most profitably. Usually, but not always, that profitable option is going to be the most sustainable one. To be sure, not every recycler is an environmentalist, and not every recycling facility is the sort of place you'd want to take kindergartners for a field trip. But in an age of conspicuous consumption, the global recycling business has taken on the burden of cleaning up what you don't want, and turning it into something you can't wait to buy.

In the pages to follow I'll tell the story of how the very simplest of human activities—reusing an object—evolved into an international business that has played a key role in the globalization of the world economy over the last three decades. It's a murky story, obscure even to those who care very much about what happens to what they toss in their recycling bins. Like most stories that are at least partly hidden from view, the story of globalized recycling reveals uncomfortable truths and the singular, sometimes brilliant, characters who grapple with them on our behalf.

Most of those characters, like Raymond Li, share a talent for spotting value in what others throw away. In colonial-era America, Paul Revere demonstrated that talent, smartly buying scrap metal from his neighbors for remelting in his blacksmith shop. In late 1950s America, that talent was applied to finding a way to make a living by recycling the tens of millions of automobiles abandoned across the American country-side. Today, it's a talent being applied to recycling the rare and valuable elements buried inside the smartphones, computers, and other high-tech devices that middle-class people throw away like candy wrappers. More often than not, though, the genius is commercial, not technical. Today recycling is as risky and rewarding as any global business, if not more so. Huge, mind-bending, Silicon Valley–scale fortunes have been built by figuring out how to move the scrap newspapers in your recycling bin to the country where they're most in demand.

Of course, for most Americans and other people living in wealthy de-veloped countries, recycling is an environmental imperative, not a busi-ness. From that perspective, recycling consumes fewer trees, digs fewer holes, and consumes less energy than manufacturing from virgin mate-rials (a recycled beer can requires 92 percent less energy to manufacture than one made from virgin ore). But without financial incentives, no ethical system is going to transform an old beer can into a new one.

The global recycling business, no matter how sustainable or green, is 100 percent dependent upon consumers consuming goods made from other goods. This unbreakable bond—between raw material demand, consumption, and recycling—is one of the dominant themes of the pages to follow. The calculus is simple: the only reason you can recycle is be-cause you've consumed, and the only reason you can consume certain products is because somebody else recycled. Around the world, we recy-cle what we buy, and we buy a lot.

Nonetheless, despite what some recycling companies will tell you, many goods—such as smartphones—are only partially recyclable, and some—like paper—can only be recycled a finite number of times. In that sense, recycling is just a means to stave off the trash man for a little lon-ger. If your first priority is the environment, recycling is merely the third-best option in the well-known pyramid that every American schoolchild learns: reduce, reuse, recycle. Alas, most people have very little interest in reducing their consumption or reusing their goods. So recycling, all things considered, is the worst best solution.

But what a solution! According to the Institute of Scrap Recycling Industries (ISRI), a Washington, DC–based trade group, in 2012 the 46.35 million tons of paper and cardboard recycled in the United States saved 1.53 billion cubic yards of landfill space; the 75.19 million tons of recycled iron and steel saved 188 billion pounds of iron ore and 105 billion pounds of coal (roughly 60 percent of American steel comes from scrap metal); the 5.45 million tons of recycled aluminum saved more than 76 million megawatt hours of electricity. In China, where industry is far more polluting than in the United States, the numbers are even more astonishing, and arguably more important. According to the China Nonferrous Metals Industry Association, recycling of metals (not including iron and steel) between 2001 and 2011 saved China 110 million tons of coal and the need to excavate 9 billion tons of ore. During that same decade, China's devotion to recycled aluminum prevented 552 million tons of carbon dioxide from being released into the country's notoriously polluted skies. Today, China is the world's biggest consumer of copper, and fully 50 percent of its copper needs come from recycling. Wherever there's a recycling industry—and it's everywhere—there are examples like these covering every type of recyclable good, clothing to car batteries.

If this book succeeds, it won't necessarily convince you to embrace the oft-gritty reality of the recycling industry, but it will certainly help you understand why junkyards look like they look, and why that's not such a bad thing. In my experience, the worst, dirtiest recycling is still better than the very best clear-cut forest or the most up-to-date open-pit mine.

Notably, there are no blue or green recycling bins at Raymond Li's Yong Chang Processing, no posters encouraging people to "Reduce, Reuse, Recycle," no cardboard boxes filled with used office paper next to the copy machine. It's a tough factory in a tough industrial town cut from ancient farm fields and staffed by migrant laborers looking for a better life. Superficially, at least, it doesn't seem to have much to do with the neatly sorted cans, bottles, and newspapers that so many Americans set onto curbsides, or carefully sort in the trash rooms of their apartments, co-ops, and condos.

It's important to keep in mind that Raymond Li's success isn't about exploitation, any more than an American junkyard exploits its employees. Rather, Raymond Li is an opportunist who long ago recognized a simple fact: China's development into what will soon be the world's largest economy created an appetite that can only be filled by importing scrap metal, paper, and plastic. If China didn't import those resources, it'd have to dig and drill for them.

As I stand in Raymond Li's factory, watching his employees mine copper from Christmas tree lights, the question that immediately comes to mind is: Why can't somebody recycle Christmas tree lights in the United States?

The reason, as I've learned over a decade visiting recycling facilities all over the world, isn't technology (Raymond Li's water table is just a fancier version of the pan that gold prospectors once used to separate gold nuggets from gravel). Rather, the issue is business: as of 2012, fast-growing China accounted for 43.1 percent of total global copper demand. Meanwhile, the slow-growth United States accounted for only 8.5 percent. That's the difference between a country (China) that has a growing middle class and lots of buildings and infrastructure yet to build, and one (the United States) where incomes have stagnated and infrastructure spending peaked decades ago. If you're building a copper factory somewhere in the world these days, it's likely in China. If you're building a recycling plant to feed that copper factory, it might as well be in Shijiao.

But that doesn't mean there's no hope for recycling in the United States. In fact, U.S. manufacturers (second only to China in total output) still use roughly two-thirds of the recycled materials that are generated within U.S. borders. The problem, if you care to view it as a problem, is that Americans don't just buy U.S.-made products; they also import vast amounts of manufactured merchandise. The result is an American economy that consumes—and throws away—much more than what is manufactured at home. That excess recyclable waste has to go somewhere. Export is one option, the landfill another. Thus it should come as no surprise to anyone that China is both the largest exporter of new goods to the United States and the largest importer of American recycling.

The story told here explains how China became America's recycling export destination of choice, and why that's mostly a good thing for the environment. After all, China and other developing countries are willing and able to recycle what the American recycling industry won't—or can't—recycle on its own (Christmas tree lights are just one minor example). When China stops buying American recyclables, those recyclables start to flow to landfills; it happened on a large scale in 2008, when Chinese factories shut down in the wake of the global financial crisis.

As a result, much of this book takes place in the United States and China. But not exclusively so: the global recycling industry is truly global, and so the narrative to follow touches on many countries, especially in the developing world.

The recycling and reclamation industry pre-dates globalization; indeed, it's as old as the first time somebody beat a sword into a plowshare—and then tried to sell the plowshare. One reason is that recycling is easy, a business that anyone can do. In the developing world, recycling a bottle or can from a waste bin is one of the few entrepreneurial opportunities available to people without capital. The negative consequences of that industry—pollution, threats to health and safety—are real, but compared to the alternatives—a return to subsistence farming, an inability to pay school fees—are often accepted as fair if unpleasant trade-offs. For recyclers in the wealthy developed world, these kinds of trade-offs are unimaginable; but in India, in southern China, in the lower-income parts of Los Angeles, they're far less important than the pursuit of good nutrition, safe food, clean air, and clean water. Under such circumstances, recycling someone else's garbage isn't always the worst thing. In the pages to follow, I'll explore those trade-offs.

The recycling business covers as many sectors as there are things that people consume and throw away. Over the course of a decade reporting on the industry, I've visited businesses devoted to the buying, selling, and recycling of metal, paper, plastic, oil, and textiles. I've also had the chance to visit some of the most advanced recycling facilities in the world as well as some of the most primitive, many of which are devoted to the refurbishing and recycling of specific products ranging from automobiles to televisions, Japanese pachinko machines to Indian schoolbooks.

This book will touch on all of these sectors, but its primary focus will be on scrap metal. I chose this sector for several reasons, but the most important is that the world's most recycled product (by weight) isn't a newspaper, a notebook computer, or a plastic water bottle—it's an American automobile, most of which is metal. In 2013, the United States recycled nearly 11.6 million cars (a down year, due to a weak economy and Americans holding on to their cars longer), generating millions of tons of metal that was quickly and efficiently recycled into a range of new products (mostly parts for new automobiles) around the world. Unlike newspapers, Coke cans, and computers, automobiles rarely end up in landfills. Instead, they almost always end up in recycling facilities, giving automobiles a nearly 100 percent recycling rate—something that no other product approaches (for example, in the United States and Europe a relatively paltry 65 percent of paper and cardboard is recycled).

That wasn't always the case. As I'll document in this book's later chapters, only fifty years ago, automobiles were almost impossible to recycle, and as a result millions of abandoned car bodies cluttered and polluted American cities and the U.S. countryside. Collectively, they formed one of the most serious environmental crises in the United States—and then, due to a scrapyard innovation, the problem was solved. Today, the methods and means by which the United States solved its abandoned car problem are being adopted by China and other developing countries with eager car buyers.

My focus on scrap metal is also intended to expand the discussion of recycling beyond the home recycling bin, all the while emphasizing that the means and markets used to recycle the contents of the home recycling bin are the same ones that apply to the sloppy jalopy being towed into a local junkyard. In fact, statistically speaking, the volume of recyclables harvested in American homes and offices is often just a small percentage of what's harvested in total in the United States.

Take, for example, aluminum. In 2012, according to the U.S. Environmental Protection Agency's most recent data, Americans harvested 710,000 tons of aluminum from their home and office waste, most in the form of beer and soda cans. That might sound like a lot—and it is—but in fact it's just 26.1 percent of the total aluminum scrap harvested in the United States that year! The rest—the other 2.72 million tons, according to data from ISRI—was harvested from factories, mines, and farms; from power lines, automobiles, old machines, and countless other sources that have nothing to do with home or office recycling bins. To understand why and where the aluminum in your recycling bin goes, you'll need to understand where all of that other aluminum goes, too.

Finally, I focus on metal for a personal reason: I am the son of an American scrapyard owner. That business (still operating, modestly, in north Minneapolis) and this industry (operating all over the world) inform my outlook on life in fundamental ways that I'll discuss throughout this book. The story you are about to read is, in part, my story, an adventure story, whereby a small-time scrapyard kid left home and hitched a ride on all the junk his family was sending to Asia.

That last point underlines something that should already be obvious: I love what my grandmother called "the junk business." Some of my earliest and happiest memories are of wandering among the family junk inventory, often with my grandmother, finding treasures. When I'm on vacation, and there's a scrapyard that I can visit, I'll usually go (apologies to my wife). When I visit scrapyards, whether in Bangalore, Shanghai, or São Paulo, I know I'm at home. Believe me, I'm aware of the industry's faults, and they'll be explored in these pages. But for all of its problems—and they are rife—the world would be a dirtier and less interesting place without junkyards.

CHAPTER 1

Making Soup

Here's something true in all places and times: the richer you are, and the more educated you are, the more stuff you will throw away. In the United States, wealthy people not only buy more stuff but they buy more recyclable stuff, like the recyclable cans, bottles, and boxes that contain the goods they covet. That's why, if you take a drive through a high-income, highly educated neighborhood on recycling day, you'll see green and blue bins overflowing with neatly sorted newspapers, iPad boxes, wine bottles, and Diet Coke cans. Meanwhile, take a drive through a poor neighborhood, and you'll invariably see fewer bins, and fewer recyclables.

The people in the wealthier neighborhoods who did that sorting, that harvesting, were good stewards of their trash. But they wouldn't have had the chance to be good stewards if they weren't also very good consumers of stuff (just as poor people don't harvest as much recycling in part because they don't buy as much). There's statistical support for the observation: between 1960 and 2012 (the most recent date for which the U.S. Environmental Protection Agency provides data) the volume of recyclables that Americans harvested from their homes and workplaces rose from 5.6 to 65 million tons. Yet during that same period the total volume of waste generated tripled, from 88.1 to 250.9 million tons. No doubt Americans were doing a better job of recycling their waste, but they were also doing an equally fine job of generating it. The more numerous and wealthier they became—and the period from 1960 to 2010 was a period of intense wealth accumulation—the more waste they generated. In fact, over the course of the last five decades, the only significant annual decline in total generated waste occurred in the wake of the 2008 financial crisis and recession.

The correlation between income and recycling has been well established for decades. Consider, for example, Hennepin County, Minnesota, population 1.168 million. I was born in Minneapolis, Hennepin's largest city, and as of 2010, my hometown's recycling rate ranked thirty-sixth out of forty-one Hennepin County communities, with an average annual household recycling harvest of 388 pounds. Meanwhile, west of Minneapolis, households in the highly affluent lakeside community of Minnetonka Beach had an annual household recycling harvest of 838 pounds, putting them atop the county rankings. Why? One reason is that the median household income in Minnetonka Beach was $168,868 in 2010, compared to $45,838 for Minneapolis, a city with large pockets of poverty. Sure, there are other factors at play (at the time the data was taken, Minneapolis required residents to sort recyclables into an irritating and time-consuming seven different categories, while Minnetonka Beach required only one), but it's hard to escape the fact that places like Minnetonka Beach generate many more neat white recyclable iPad boxes, and Sunday editions of the *New York Times*, than the housing projects of Minneapolis.

Back when I lived in the United States, I had blue bins and green bins, and I felt an ethical compulsion to fill them—and, if possible, fill them more than I filled the trash bin. The paper went into one, and everything else went into the other. Then I'd drop them at the curb, but—owing to a childhood spent in my family's scrapyard—I felt as if I'd just cheated myself. Aluminum cans, I knew, were priced by the pound; during summer breaks from school I'd often be the person assigned to weigh the ones dropped off by bums, college students, and thrifty home recyclers at our family business. In her later years, my grandmother—raised in a depression-era household that saw value in everything reusable—would still insist on driving her modest number of cans to the business, rather than give them away to a municipal recycling program for free.

More often than not, in the United States and the rest of the developed world, the people who have to figure out what to do with the trash that we toss out of our homes are not teenagers on can machines but cities and a handful of large corporate waste-handling companies. In some cases they have no choice but to take what's tossed into those bins. If they had a choice, they'd take only the stuff that they can sell for a profit—like the cans that my grandmother liked to deny them. Those things that can be sold for a profit are generally things that can be easily remade into something new. An aluminum can is easy to remake into a new aluminum can; a leather suitcase, however, is hard to remake into anything at all.

Occasionally, when I drive through an American neighborhood on recycling day, I'll notice bins filled with things like old luggage, placed there out of a misplaced but righteous conviction that the companies need to do the right thing and "recycle" them too—whatever that means. But the recycling companies aren't resisting the chance to do the right thing. They just haven't found a profitable way to separate, for example, the plastic that constitutes the luggage handle from the different kind of plastic that constitutes the luggage itself. That sort of work has to be done by people who can see a profit in it, and so far the large-scale recycling companies that pick up blue and green bins haven't figured it out. But what they have started to figure out is how to dig deeper into your trash to get at the stuff that can be recycled profitably. It's not the most glamorous work, and it's generally not the sort of thing that politicians and environmentalists discuss when they discuss "green jobs." But for the right person, it's an opportunity as endless as anything dreamed up by Silicon Valley.

Alan Bachrach is the right kind of person. As director of recycling for the South Texas region of Waste Management Corporation, North America's largest recycler of household waste, he has a professional, profit-driven interest in recycling. Like so many of his peers in the global scrap and recycling trade, he has a youthfulness to him that belies his late middle age—a youthfulness that suggests nothing so much as that he really, really enjoys machines that sort trash. If there are those who feel shame for working in an industry that handles other people's waste, Alan Bachrach isn't one of them. He loves it.

We meet early in January 2012 in the visitor's area of Waste Management's $15 million new Walmart-sized recycling plant. Bachrach played a major role in designing the facility, and he's now responsible for running it. But even though we're *kind of* having a conversation, Bachrach's eyes don't focus on me, but rather on what's on the other side of a plate-glass window and two stories down: swiftly moving Class A rapids of plastic bottles, cardboard, and paper, riding up and down conveyors, over and under, around and around, until they emerge as perfectly sorted hay-bale-sized blocks of bottles, cardboard, and paper, tied together with steel ties. "You either love it or you hate it," he tells me about those who work in the industry. "You're either gone after six weeks, probably before six weeks, or you don't ever leave."

In a sense this is Green Heaven, the place where all that home recycling set out on recycling day—the paper, bottles, and cans lovingly harvested—eventually ends up. Alan Bachrach isn't exactly Saint Peter at the gates, but he's definitely in the chain of command. But then if this, the Houston Material Recovery Facility, is Green Heaven, then it must be said that Houston itself is a kind of Green Hell—at least if you care about what happens to residential waste and recycling.

The numbers tell the story. In 2012 the United States recycled approximately 34 percent of its "municipal solid waste." That is, 34 percent of the waste generated by homes, schools, and office-based businesses (but not industrial facilities, construction sites, farms, and mines) was diverted from landfills into some kind of facility that sent it on its way to a reusable afterlife. Give or take a few percentage points, that 34 percent is roughly the same percentage achieved by New York, Minneapolis, and other U.S. cities with long-standing recycling programs. But Houston? As recently as 2008 Houston only managed to recycle 2.6 percent of its municipal solid waste. The other 97.4 percent? By and large, it was landfilled. Since 2008, the rate has been pushed up to "six or seven percent," according to a sheepish Alan. That's not good, by any definition. How to explain it?

For people who live in places like San Francisco, where the recycling rate exceeds 70 percent, a popular explanation is that rednecks don't like recycling. But that's not only condescending, it demonstrates a profound misunderstanding of how and why San Francisco's trash is recycled at such a high rate.

No doubt culture, education, and income play a role in how much actual waste a particular person or place recycles. But in my experience, no culture encourages a high recycling *rate* quite like the culture of poverty. In essence: if you can afford very little, you'll tend to reuse a lot. So in San Francisco a glass jar of Trader Joe's bruschetta is likely headed directly to the recycling bin; in the slums of Mumbai that same jar—if somebody could afford it—might very well become a kitchen implement. The slum dwellers of Mumbai have a far higher recycling rate than the suburbs of San Francisco because (a) they consume less—for example, no iPad boxes to recycle—and (b) daily survival requires thriftiness. But no matter how poor or eco-conscious a particular population is, the degree to which they recycle primarily comes down to whether or not someone can derive some economic benefit from reusing waste. In Mumbai the benefit is largely a matter of personal economy; in wealthy San Francisco, where few residents worry about the pennies they might generate from a pile of newspapers, it's a recycling company that has to find an answer to the question of whether there's economic benefit in picking up someone else's waste.

Houstonians, like most Americans, don't share an interest in practicing Mumbai-style thriftiness. So that places pressure on recycling companies, who unfortunately have found it very, very difficult to be profitable in Houston. The problems are several. First, Houston is big, but its population density is very low—around 3,300 people per square mile. San Francisco, by contrast, has a population density that exceeds 17,000 people per square mile. From a demographic standpoint, that means there will be more recycling bins per square mile in San Francisco—because there will be more households per square mile—than in Houston. What that means, from a recycling business standpoint, is that a recycling truck has to drive much farther to pick up, say, a thousand pounds of newspaper in Houston than it does to pick up the same weight of newspaper in San Francisco. In other words: a Houston recycling company has to work harder, and pay more money, for the same revenue as a San Francisco recycling company.

One way to overcome this problem is for local governments to subsidize recycling—and some do. But in tax- and fee-averse Houston, that's a tough proposition, especially because Texas has some of the cheapest landfill rates in the United States. Reasonable taxpayers—not to mention politicians—might ask why they're being asked to pay more to recycle, when the same trash can be landfilled for so much less.

The other way to overcome the problem is to encourage Houston's households to harvest more recyclables so that each pickup is potentially richer for the recycling companies. Believe it or not, that's really easy to do (and to do without encouraging an increase in consumption). Here's how: take away the two, three, and sometimes seven bins into which some American households are expected to sort waste recyclables, and replace them with one big bin where everything recyclable can be dumped. This is called single-stream recycling (as opposed to dual-stream recycling, which requires one bin for paper and another for everything else). In communities where this has been tried, recycling rates have increased by as much as 30 percent. And why not? Like it or not, even eco-conscious people are sometimes too busy to be bothered with the need to separate their trash into multiple containers ("playing with garbage" is how I like to describe it). So Waste Management has spent the last several years rolling out single-stream recycling in Houston.

But if Houston's households aren't sorting all of the extra recycling they're dumping into Waste Management's trucks, how is Waste Management supposed to extract more recycling from it? That's where Alan Bachrach, a bunch of engineers, and $15 million comes in.

In high school, some kids look for jobs at McDonald's, and some kids are happy mowing lawns. Alan Bachrach wasn't that kind of kid. Rather, he was the entrepreneurial kind, the sort who looked for things that could be sold for more than they cost to buy. He found two: the computer punch cards that were, until the late 1960s, the primary means of feeding data into mainframes, and continuous-feed computer paper. Both were 100 percent salable, for cash, to local paper scrapyards, where they were prepared for processing into new paper. Thus, Alan Bachrach had plenty of pocket money in high school. In fact, I'm guessing he had more money than most.

"I got very lucky," he tells me when I ask about how he was attracted to the recycling business. "It fit my aptitudes and my ADD and my OCD very well." As with many young entrepreneurs who find their calling early, Alan's college career didn't last long, and after dropping out he went to work for a friend's trash-hauling company. There he introduced the trash men to the revenue potential of selling recyclable paper and cardboard to scrapyards, and for the next three decades he devoted his life to recycling paper and cardboard generated in Houston-area businesses (not homes). Everything changed, however, in 2008, when Waste Management, in search of recycling companies that could help it establish a residential recycling business in Houston, decided that Gulf Coast Recycling—the company where Alan had spent nearly three decades—was the one to help them do it. This was well timed. Alan wanted Gulf Coast to move into residential recycling, but they lacked access to large volumes of recyclables. "Those are collected by trash companies," he explains. "And so it's very difficult to justify fifteen, twenty million dollars of equipment when you don't have the feed materials secured."

"You need scale," I respond.

Standing beside him, Lynn Brown, Waste Management's VP for communications, pipes up: "Or you need a municipal contract in the city of Houston."

Alan smiles widely. "Scale is very important in this business."

Waste Management acquired Gulf Coast Recycling in 2008, and in 2010 it began to transform this GCR facility into a single-stream recycling plant that opened in February 2011. Today it sorts between 600,000 and 700,000 pounds of single-stream recycling per day. That's roughly the weight of an Airbus A380 jet—measured out in newspapers, plastic milk jugs, beer cans, and shoe boxes. When I ask for an estimate of just how many households those pounds represent, Alan tells me that, on average, a Houston family generates fifty pounds of single-stream recycling per month. However, not everybody recycles, not everybody rolls out their container on a weekly basis, and some—like Alan's family, which rolls out six (!) containers per week—recycle more than the average. At the same time, a small percentage of the material handled at this facility continues to come from commercial sites, such as the Dumpsters filled with cardboard behind supermarkets. Still, a rough calculation suggests that—on a daily basis—the Houston Material Recovery Facility processes a volume equivalent to the monthly recycling generated by approximately 12,000 Houston households.

"You ready to take a walk?" Alan asks me with that childlike gleam. We're accompanied by Matt Coz, the Waste Management VP in charge of growth and commodity sales—that is, making money off the stuff processed at this plant—and Lynn Brown. Both of them have been through the plant many times—Matt was intimately involved in its planning—but I don't sense any weariness at the thought of touring it again.

The four of us walk outside and around the building, to an enclosed receiving area where a truck is tipping a load of recycling onto the concrete floor. Single-stream recycling hisses more than it clanks, mostly due to the fact that 70 percent of it is paper—junk mail, newspaper, office paper. A front-end loader of the sort most people are accustomed to seeing dig in the dirt at construction sites rolls up and digs into this mass of well-intentioned waste, picks it up, and dumps it into what Alan tells me is a device that feeds the stuff onto conveyors at a steady, uniform pace. "That's really important," Alan says, "if everything you're about to see is going to work properly and consistently."

We enter into the Walmart-sized space that I saw from above, and I swear, the first thing that comes to mind is Willy Wonka's chocolate factory: conveyors of trash rush upward and release their cargo into spinning stars that toss it about in a manner that I can only describe as joyful, like popcorn jumping in a frying pan. Some continues along, some drops away. I see detergent and shampoo bottles zipping by at a pace that exceeds 400 feet per second (Alan asks me to keep the actual speed to myself—it's a trade secret), and I see milk bottles dropping, from points unknown, into a giant cage. Something occurs to me: "Kids would love this place," I call out to him, but he doesn't respond, perhaps because (a) it's so obvious; or (b) he can't hear me because my voice is completely lost in the roar of machines, the hiss of paper, and the crash and bang of glass, aluminum, and plastic.

We climb a stairway to what Alan calls the "pre-sort." Here two workers stand over a high-speed conveyor that carries freshly arrived, unsorted "recycling" that needs to be, well, recycled! One of them reaches out and grabs a brown plastic bag from the blur, and just as quickly it disappears, sucked up by a large vacuum tube positioned directly above them, all Willy Wonka–like. Then he does it again! "Not everybody can hack this job," Alan leans over to say, nodding at the speeding, blurry line. "Some people get dizzy, throw up."

That's not what interests me, however. "What happens to the plastic bags?"

"Plastic bags are the worst," Alan calls out to me. "They get tied up in the axles and we spend hours pulling them out."

I make a mental note to myself: never again use plastic bags to contain my old beer cans. "But can you still recycle them?"

"Sure!"

One of the sorters grabs a hunk of something—it happens so fast I can't tell what—and drops it down a square chute that leads to—where? It could be the other side of the planet, for all I know (China?). "The other job is to pull out big pieces of plastic and trash," he adds and points me farther along to the spinning stars I saw from below. I don't have a chance to ask where that chute goes.

The conveyor feeds into the stars and newspaper bounces and froths atop them like white water on churning waves. The stars—they're made of a specialized plastic that wears well—are spaced at intervals to allow plastic, glass, and aluminum to fall down onto another line. The newspaper, meanwhile, dances right across the stars and emerges on the other end, separated. Meanwhile, below, the material that dropped through the stars, including more paper, is conveyed into yet more stars, spaced at smaller intervals that convey out even more paper—in smaller sizes—while the plastics, glass, and cans continue to fall away. It's a cascade, each step angled steeper than the last, and at each level paper and plastic are separated. It's a key process—perhaps *the* key process—in cleaning up a waste stream that's roughly 65 to 70 percent newsprint, office paper, and junk mail.

Below, aluminum cans are literally ejected from the system by a device that creates an electrical current that repulses metals. To me, it looks as if the cans are jumping to their deaths from the streams of paper and plastics, into a cage where they're collected for companies that remelt them. The glass, meanwhile, is removed by several processes that take advantage of the obvious fact that glass is heavier than paper. Think of it this way: if you place a beer bottle next to a pile of newspaper coupons, and aim a hair dryer at both, you'll likely be left with only a beer bottle. That's a rough approximation of the physics that Waste Management uses to separate the two materials.

I must admit, I'm really getting into this when, suddenly, the entire line grinds to a groaning halt. I turn to Alan. "Everything okay?"

"Something probably got stuck," he says with a wave of the hand indicating that this is common. "Probably a bad piece of material. It shuts down the whole thing."

As we wait for it to restart, I lean over a rail and realize I'm maybe twenty feet off the ground, and we've only just begun to tour this behemoth. Alan tells me that it takes a piece of recycling roughly twelve minutes to go through the entire system, beginning to end. Below, I watch a forklift race across the floor with a bale of what looks like junk mail—thousands and thousands of pieces of junk mail. That bale will be placed into a shipping container and, more likely than not, shipped to China for recycling into new paper.

With no warning, the engines start to roar, the conveyors start to run, and the giant recycling machine slowly comes back to life, one distant conveyor, star screen, shaker table, and goodness knows what else at a time. "You can't just turn it all on at once," Alan explains. "It's so complicated you have to stage it." If my ears can be trusted, it takes a good fifteen seconds or so before the conveyors are all running again.

We climb more stairs, moving ever higher into the system. There's no more paper up this high—that's all been pulled from the system. Now it's all about separating the different kinds of plastics. "This is my guys' favorite one," Alan tells me, and he nods at a yellow device that hangs over a blur of bottles. It contains two hundred sensors that shine infrared light onto the trash passing beneath them. When the light reflects off, say, a red Tide detergent bottle, nothing happens; if it reflects off a white Minute Maid orange juice bottle, nothing happens. But if, as the trash whizzes by, the light reflects off a clear Coca-Cola bottle, the computer records exactly where it is on the precisely timed conveyor.

"Hear that?" Alan asks loudly, with an impish grin.

Through the clamor, I hear irregular, sharp pops of compressed air, like tiny gun bursts. A few feet from the sensors, I see an Aquafina bottle stagger backward onto another conveyor, as if shot dead, followed by a Coke bottle—also shot dead. The computer knows exactly where the bottles are, and how long it takes for them to arrive at the air guns. I can see the nozzles, now, tiny needle tips capable of sending an empty bottle flying. The rate of fire reminds me of nothing so much as what you might hear at a crowded gun range. *Phhht phht.* Two bottles down. *Phht Phht Phht.* Three more bottles down. Pause. *Phht.* According to Alan, this one machine, its sensors and air guns, replaces six to ten manual sorters, who, unlike machines, might become fatigued and nauseated, watching the swirl of plastic beneath them.

Still, infrared light sensors, for all of their sophistication, have limitations. One of them, according to Alan, is that "they can't sort white polyethylene bottles from colored polyethylene bottles." In layman's terms, that's the process of sorting a red Tide detergent bottle from a white Minute Maid orange juice bottle. But have no fear: "We have the most

sophisticated equipment available: a person." Sure enough, three human beings stand over a conveyor, grabbing the white bottles and tossing
them down chutes. At their best, people can handle perhaps forty-five
"sorts" per minute—not bad, but certainly not nearly the hundreds that
an array of air guns and sensors can handle. The technology doesn't
exist for this type of plastics, so people will have to do.

For all of Alan's joking about people as technology, however, he never
once diminishes the dignity of the work done by his sorters. Like many
of the scrap entrepreneurs I've met over a lifetime around this industry,
he identifies with them. After all, they're all in the business of sorting
other people's waste. "These are people I'd stop bullets for," he tells me.
"These are wonderful . . ." He hesitates for a second, and then revs right
back up. "People think that because these are minimum wage jobs or
close to minimum wage jobs, it's a very high-turnover position. I've got
ten, fifteen, twenty years folks."

It may not be the highest paying job, and it may not be glamorous or
the sort of thing that your kid will want to tell his friends about. But if
you're looking for a consistent job, with benefits, where layoffs are almost unknown, then it doesn't get much more consistent than American recycling. In Houston, a town that knows the heartbreak of boom
and bust better than most areas, a stable job like that is worth more than
just salary. "My boss," Alan tells me, his smile widening again, "the guy
that runs this area for Waste Management, is in the business because his
dad was in the oil business and he watched him during his youth, boom
and bust. And he said, 'I gotta find an industry that's always steady.' "

Far away from Houston, in offices located in Denver, Milwaukee, Bos-
ton, and Chicago, thirty-five men and women arrive at work each day
tasked with finding a home for the sorted recyclables that come out of
Houston, and Waste Management's dozens of other recycling plants
throughout North America. To them, a bale of recycled detergent bottles
is no more or less virtuous than the barrel of oil from which those bottles were originally made. There's nothing sentimental in the work that
these men and women do, nothing particularly green or eco-conscious.
Their job, pure and simple, is to obtain the best price. So, for example, if
a factory in Fuyang, China, is offering a better price for those bottles,

then those bottles will very likely go to Fuyang; but if, as is quite likely, a U.S.-based manufacturer needs them more, and is willing to pay to prove it, then they'll stay in the United States. It's just business, constrained only by the cost of shipping and the laws and regulations of the United States and wherever Waste Management would like to export to.

Matt Coz, the VP for growth and commodity sales, who has accompanied Alan and me on our tour of the Houston facility, is in charge of that marketing. As my tour winds up, we stop into a room where bales of what was once recycling, and now qualifies as commodities, are stacked in columns that reach fifteen feet off the floor. "The aluminum cans," Matt says to me, pointing at a shiny bale. "They're a small percentage of the overall weight that comes in here, but a big percentage of the overall value that comes out." Put differently: a pound of newspaper is worth a couple of pennies; a pound of cans extracted from that bale is—as I stand with Matt—worth around 54 cents on the North American markets. Now imagine that you have tons of those cans, and you've paid nothing for them: that's the kind of profit that a scrap man, a recycler, is in business to achieve.

I turn to look back into the main room, where a forklift is pulling away with a bale of newspaper, and then I turn my attention back to the storage room. There I notice, stuck between bales of plastic detergent bottles, several bales of what look like large plastic Waste Management recycling bins, wheels and all. "Are those what I think they are?"

Alan laughs. "They sometimes drop into the trucks when the drivers are trying to empty them. Happens all of the time."

I walk closer to one of the bales. The recycling bins look perfectly good, but for the fact that they've effectively been flattened into two dimensions. They're trapped among two-dimensional buckets, laundry baskets, and a few milk bottles, like a dinosaur skeleton amid fossil clam shells. I recall that in 2008 a group of San Francisco writers— including Jonathan Franzen—held a fund-raiser to buy those nonrecycling hicks in Houston a gift of 276 recycling bins. It was a well-intentioned if somewhat condescending idea, but I'm pretty sure that—if what I'm seeing before me is representative—those bins were long ago flattened, shipped to China, and made into laundry baskets for Shanghai's up-and-coming middle class. Practicality and profits nearly always trump good intentions, condescending or not, in the global scrap business. "You don't pull them out?" I ask.

"It's not worth it," Alan answers. "By the time they arrive here they've been smashed down in the truck. Doesn't pay to shut down the machine, pull them out, and drive them to someone's house."

This Houston recycling line is one of the newest and most advanced facilities of its kind in Waste Management's growing arsenal of thirty-six single-stream recycling lines. That means, in all likelihood, it's one of the most technologically advanced household waste sorting lines in the world. Still, for all of the things that the Houston facility is, what it's not—and this is very, very important—is static. It can't be: the consuming habits of Houstonians are constantly changing, and thus so is what they waste. This single-stream line will evolve with those changes.

Today, the line is designed and tuned to receive a single stream that's roughly 70 percent newsprint, magazines, and junk mail. But that's changing: more and more Houstonians are reading their papers on e-readers. There are good statistics to back up the shift: according to Moore & Associates, an Atlanta paper recycling consultancy, in 2002 Americans recycled 10.492 million tons of newsprint. In 2013 they recycled 5.631 million tons. As a direct result, junk mail, as a percentage of the recyclable waste stream, is growing. For the purposes of a mechanical sorting line, that's a big shift: junk mail weighs less, and is worth less, than newsprint. In time, perhaps when newspaper drops another 5 percent as an overall percentage of the waste stream, Alan and Waste Management's engineers will have to tweak the machine—maybe add a star screen or two, maybe change the speed of some conveyors—to keep up with the times.

"I like to think of it as making soup," Alan tells me. "Soup can handle a little pepper, a little garlic, but you just can't make pepper and garlic soup. It's a constant mixing, it really is a craft." "It" is sorting what Americans throw away, and if Americans can't be bothered to do it themselves, then Alan and Waste Management's engineers are more than happy to create a multimillion-dollar technology to do it for them. "Those of us who love this part of the business love it because most of us are ADD and need it to be changing," he tells me. "You'd kinda get bored if it were static."

*   *   *

As it happens, demand for newsprint is growing in China, too, and so is the junk mail. But you won't find a Chinese version of Alan Bachrach making soup, just as you won't find an Indian, Kenyan, Vietnamese, or Jordanian one, either. The reason is that most of the world remains poor enough to justify employing people to do what Alan does with star screens and air guns: sort and harvest recycling. Likewise, if a place is too poor to justify a setup like Alan's, it's probably still too poor to generate enough recyclables to justify investing in the setup.

Consider what happens every night in the courtyard of the Shanghai high-rise where I've lived for nine years. Just past midnight you'll likely hear the clank of a bottle bouncing across concrete. If you follow that bounce back to its source, you'll come to a concrete hut, not much larger than a single-car garage, from which fragrant trash has been spilled several feet into the narrow asphalt road. It doesn't look like American trash: there are few boxes, cans, bottles, or anything else rigid that might have once contained something else. Rather, it's mostly food waste—peels, husks, bones.

Move a little closer, and you might see two or three hunched shadows atop the oozing mess, canvas bags swung over their shoulders, scrounging through it with bare hands, searching for metal cans, plastic bottles, or perhaps something better yet—a stray coin. They're not Shanghainese—the proud Shanghainese would never be caught scrounging through their neighbors' garbage, even in the middle of the night—but often families of poor migrants from farms in the less wealthy provinces, making the best living that they can. For the privilege of scrounging, I have it on good authority, they need to pay a small bribe to the attendant at my building's front desk, and agree to clean up everything by daylight. Both requirements are accepted without complaint, but the latter, in particular, is no problem: the families engaged in this kind of raw material extraction need to visit several compounds in a night if they're going to make a decent living. They need density.

China does not lack for trash to be mined by migrant families. In fact, as of 2008 or so, China generates more trash than the exceedingly wasteful United States—roughly 300 million tons per year, compared to around 250 million tons in the United States. Still, on a per capita basis the Americans have the Chinese beat four to five times over (Americans are richer). For example, Americans consume 653.62 pounds of paper

per capita per year, while Chinese consume 98.34 pounds, and Indians, on average, consume an unimaginably paltry 18.7 pounds. Even accounting for the much larger populations of the two developing countries, and the resulting larger total volume of paper, it's definitely the case that smaller per capita consumption means that households, and scrap peddlers in these countries, face a much easier task sorting recyclables.

The trend is not positive if you care about resource conservation. Chinese consumers are joining the global middle class in droves, adapting the consumption habits that go with that status. For example, between 2000 and 2008—a period of historic economic growth in China—the prepackaged food industry in China grew 10.8 percent. That shift, from buying fresh food to buying food in plastic, aluminum, and glass containers, has had a profound impact on what happens behind my apartment building in the middle of the night—and in Chinese landfills every day.

The good news, for now, is that China sends very, very little recyclable waste to landfills or incinerators. Families that pick through trash in the middle of the night are just the last screen in a profit-driven process that, if you wait outside my building through the night, begins at the gate at first light. There you'll see a squat and brawny migrant woman— she looks fifty, but she's actually in her thirties—waddle across the street wearing a fanny pack stuffed with small money, and carrying a small hand scale. If anyone is the Chinese equivalent of Alan Bachrach, presiding over a system that harvests recyclables from the trash, she is it. Her destination is the pile of cardboard tied with twine that's waited beneath the night watchman's gaze, and a waist-high balance scale of the sort that you might find in a doctor's office. As she pulls out the big scale, the early-rising old ladies in my building wander downstairs carrying a few plastic bottles, perhaps a small cardboard box or two, and maybe a small plastic bag filled with cans. The bottles and cans are priced individually; the boxes are attached to a small hook on the hand scale and weighed. Payment is equivalent to a few pennies, which the early risers take to the wet market, in search of the day's vegetables.

Morning brightens, traffic thickens, and the recycling lady is joined by her slight husband. While he handles the walk-up transactions, his wife ventures up into my building, responding to dispatches relayed through the guard booth: somebody bought a television, and wants the big box in which it was delivered removed; somebody else might've been collecting newspapers for weeks, and—at the prodding of a fed-up spouse—wants to sell them. Through the morning, she rides the elevator, up and down, paying a few pennies—the market price—for whatever's available for recycling. Then she carries it down and sorts it into the appropriate piles.

While she does, men on bicycles that pull small trailers begin to arrive. Some are there to buy newspapers, others for cans. Whatever it is they want, they pay more for it than what she paid for it in the building, then tie the goods to their trailers and pedal off. Before the end of the day, they'll sell the collections to a small scrap company, one with a warehouse rather than a street corner. But the concept is the same: buy low and sell high. At the small scrap company, the bicycle recyclers will be joined by other bicycle recyclers, each with similarly modest trailers full of scrap ready to be sold and—later—combined into bigger loads that can be sold to paper mills, aluminum smelters, and other manufacturers in need of raw materials.

There are no good statistics on how much of China's household waste is recycled, and with much of China still rural and undeveloped, collecting such statistics would be prohibitively expensive, if not impossible. But one thing that everyone agrees upon—from government officials to midnight garbage scroungers—is that by the time a load of Chinese trash arrives at a landfill, very little that's reusable or recyclable is left in it. Houston and San Francisco would be very glad to say the same.

And yet: Shanghai doesn't distribute recycling bins to its residents; it doesn't have a local equivalent to the multimillion-dollar Houston Material Recovery Facility; it doesn't have infrared sensors that machine-gun plastic bottles off speeding conveyors. Rather, it has hundreds of thousands of small businesspeople who make their livings by buying cardboard, paper, and cans from millions of people who would never give away perfectly reusable material for free. Whatever is left over is scrounged, completely. There's no need for single-stream schemes to raise the recycling rate in Chinese cities because, in the end, my Chinese neighbors have something that most Americans don't: a recognition that "the recycling" is worth more than virtue.

It's worth money.

CHAPTER 2

Grubbing

It isn't just the Chinese who think of recycling in financial rather than strictly virtuous terms. The world's largest recycling industry—the U.S. one—was also born from self-interested motives, and for more than a century it thrived and recycled with little notice until, in the 1960s, a nascent American environmental movement rebranded the business.

The environmentalists certainly had and have a point. Resource scarcity is a serious problem that will only grow in importance as the developing world—China to India to Brazil—achieves the means to enjoy American-style middle-class consumption. So, barring great advances in the mining of the moon and other extraterrestrial bodies, the next best option is to reuse what we already have.

Reuse and recycling, however, isn't always easy. It requires ingenuity, and it requires entrepreneurship. These days, those qualities are most often seen in developing Asia, where consumption rates are increasing rapidly, along with all of the recyclable waste that entails. In my experience, a desire to save the earth isn't high on the list of the Asians who are trying to figure out how to make money off the growing market for recycling. But that's nothing new: in the United States, the entrepreneurs who invented the global recycling industry weren't inspired by philanthropic motives, either.

These days, the captains of American waste and recycling occupy boardrooms and BMWs. But the businesses over which they preside weren't started in the boardrooms. Rather, they were started with backpacks, pickup trucks, and perhaps a discrete backyard or two. Only in time, after years of growth, were they big enough to acquire boardrooms—or be acquired by boardrooms. In every case, though, the spur to growth was the same: somebody was short a resource, and somebody else with some ingenuity and entrepreneurship had an idea for how to provide it.

Seated in his office, Leonard Fritz is striking to behold: late-morning light comes through the windows and reflects off a nearly full head of white hair combed back into a pompadour. He wears white slacks, a white short-sleeved shirt, and a white T-shirt beneath it. The only shocks of color are a gold chain around his neck, and the large amber-colored sunglasses that wrap around long features softened into kindness by work and age. He's not physically imposing, but those shoulders are still broad, and it's not hard to imagine the steely body that once supported them. When he shakes my hand, I notice the strength of his work-scarred grip.

Over his shoulder I can see the skeleton frame of an abandoned steel mill, dark on the horizon, out the window. He recalls freelancing there as a young man, digging in its dumps for the fragments of manufacturing that the mill—a supplier to the then-booming Detroit auto industry—considered waste. Now, in his eighties, he spends most of his working hours in the second-floor office of this two-story building that once belonged to another steel mill, and now belongs to the company he founded when he was—more or less—nine years old. It's an extraordinary enterprise: in 2007, the biggest and most profitable year in the history of the U.S. scrap industry, his Huron Valley Steel Corporation received over 1 *billion* pounds of scrap materials for recycling. In 2011 it processed 775 million pounds. Few recycling businesses in the world have ever approached that scale.

"I was born in 1922," he tells me with a rapid, stutter-stop delivery. "Columbus's birthday, October the twelfth. We were in the midst of a depression." His mother, he tells me, worked in the scrap rag trade, sorting old household and industrial-use rags into those that could be washed and reused and those that would be shredded into fiber to be used in paper. During the first half of the twentieth century the rag trade was so big that it had its own trade journals, and served as a last-resort industry for those unable to do much else. These days it's all but forgotten except by those who had a relationship with it. "Two dollars per week," Leonard recalls. "Five, six cents per hour."

Recycling wasn't anything special in those days. In fact, the word *recycling* had only just been invented. According to the *Oxford English Dictionary*, it came into existence in the 1920s, when oil companies needed a word to describe how they recirculated unrefined petroleum through a refining plant to reduce impurities. It was recycling of a type, but certainly not the type that the word suggests today. Another half century would have to pass before the term for recycling oil would become synonymous with the civic-minded act of collecting newspapers and cans to remake them into new objects.

What we now know as recycling, Leonard Fritz and his family knew as "grubbing." It was what you did when you couldn't do anything else. Thus it happened that nine-year-old Leonard, desperate for new school clothes, went to work in the summer of 1931, "grubbing" in the dumps on the outskirts of Detroit. They were not, he makes clear to me, "aristocratic" dumps, but rather dumps for poor people.

"At the base of the dump they'd have like a hobo village," he explains. "Old tar-paper shacks. Fifty-gallon drums for a heater, stuff like that. There weren't any other kids out there . . ." The dump itself was located in a pit, beneath a ledge where the garbage trucks tipped their loads. As Leonard explains it, when the trucks arrived, everyone—hobos and nine-year-olds alike—gathered, ready to jump on anything that could be resold. Bottles were 3 cents per dozen, but the real prize was a Roman Cleanser bottle that, Leonard recalls clearly, was worth 5 cents. "That would cause a fight over there," Leonard sighs. The hobos, as he describes them, showed up at the dump armed with mop handles minus the mop. The exposed hook where the mop itself would be affixed served as a weapon that could hook a hand. "They didn't give a damn what age you were, who you were. It was like blood money."

Nobody grubs in American trash dumps anymore (instead, they sort bottles into blue and green bins). But they did within Leonard Fritz's life-time, and they still do across the developing world. I've seen dumps like Leonard's in India, Brazil, China, and Jordan, salted with impoverished people, often mothers and children, literally scraping at a living. The most famous is surely in Mumbai, movingly depicted in the film *Slumdog Millionaire*, its orphans grubbing in search of just enough recyclable trash to exchange for food. I've seen the fights that Leonard describes, but only between children. Adults usually grub separately, and most everybody grubs for subsistence.

In 1931 Leonard was grubbing for school clothes and a vaguely defined future. By that measure, he was better off than most anyone else at the dump where he worked, and by the end of the summer, he had $12.45, "which was a hell of a lot of money for the summer's work back then for a kid."

So how do people go from grubbing to running actual scrap businesses? People who have better options don't generally choose to spend their time sorting through other people's garbage. It's a profession for outsiders, as pointed out by Carl Zimring, a historian of the American scrap trade, who describes the ease of joining the nineteenth-century scrap industry in a classic history, *Cash for Your Trash*. "Little investment capital was needed to enter the scrap trade. Since the work was dirty, dangerous, and low status, few natives with other prospects chose to perform it for any length of time. The low starting costs, combined with a lack of competition from established natives, made it possible for immigrants to gain footholds in the trade."

According to Zimring's work with U.S. Census data, in 1880 over 70 percent of the workers in the American waste trades were European-born, with the majority coming from Ireland, Poland, and Germany. Of those, the vast majority were Eastern European Jewish immigrants who, in addition to the facts that Zimring lists, found themselves barred from other trades due to anti-Semitism. Many settled in East Coast cities, including New York. In fact, according to one survey, by 1900 24.5 percent of New York's Jews were active in some facet of the junk trade. My great-grandfather, Abe Leder, arrived in Galveston, Texas, via Russia in the early twentieth century, but his experiences didn't differ from his late-nineteenth-century East Coast Jewish recycling forebears as described by Zimring. Barred by education and ethnicity from joining other professions in notoriously anti-Semitic Minneapolis, he started collecting the stuff that the more established locals didn't want in their homes. It could be a very worthwhile endeavor: my grandmother, the second of Abe's five children, liked to recall that her older brother Mort's bar mitzvah was paid for by the labor that she and her siblings expended "cleaning" iron from scrap copper plumbing fixtures. "That was my father's bank," she would tell me with a smirk. "And we cleaned that bank on the basement steps."

Zimring's description of what the scrap trade required in nineteenth-century America would have sounded very familiar to my grandmother, to her father—and to Leonard Fritz: "Association with filth precluded many natives with other prospects from entering the waste trade, for identifying and collecting usable scrap was not a simple or pleasant task. Successful scrap collection involved sorting through piles of refuse and having the ability to discern valuable material. The work required uncomfortable physical labor in city dumps, mills' discard piles, and other settings considered unhealthful and unsanitary."

Zimring wrote his history of the American scrap trade in the past tense, but he very well could have been speaking about twenty-first-century China and the millions of scrap peddlers who work its big cities, driving large, slow-moving tricycles retrofitted with small trailers. To the eyes of outsiders, they must seem disorganized and randomly distributed. But stop one, as I did in late September 2011 on a random street in Beijing, and you'll learn something. "The garbage trade is controlled by the people from Sichuan Province," said a wrinkled man in a plaid fleece jacket as he paused on his bicycle. Sichuan, one thousand miles and just as many subcultures away, has its own language and its own cuisine, and its migrant laborers in Beijing are as much immigrants as my Russian great-grandfather was in Minneapolis. "The recycling peddlers are all from Henan Province," the scrap peddler added, referring to another distant province, also with its own language. "And the traders who buy all our stuff are from Hebei Province."

Standing next to me was Chen Liwen, a young researcher with the Green Beagle, a Beijing environmental nonprofit, who spends much of her time trying to figure out ways to raise environmental consciousness in Beijing. "What about the Beijing people? Aren't they interested in recycling?"

The peddler glanced out at the street and a police officer who was eyeing him. "No," he answered. "The young people aren't interested in this kind of work. They're interested in big things."

As we walked away, I told Liwen about a friend of mine, also born in Hunan Province, who once recounted that when he was growing up, his parents had a very specific threat when he didn't feel like doing his homework: "You're going to grow up to be a trash picker."

Liwen shook her head. "Some of these men make good money. Better than the people who did their homework."

Needless to say, Leonard Fritz didn't do much homework. But what he lacked in book smarts, he more than gained in dump smarts.

Early in our conversation, he brought up September 12, 1931, a Saturday, and the first day of Rosh Hashanah, the Jewish New Year. He remembers it clearly because Shorty, the Jewish man who was the primary buyer of recyclables from the dump's hobos, wasn't buying that day. In the absence of Shorty's cash, the dump's "mostly alcoholic" hobos began to panic. Nine-year-old Leonard, newly enriched with the $12.45 he'd earned from a summer of grubbing and prematurely wise to the world, went off in search of his father with the question that has driven junk men since the onset of the Industrial Revolution: "If we bought that, could we sell it?"

Leonard's father said yes, and nine-year-old Leonard Fritz was on his way. "I had a very ambitious bunch of drunks who'd sell a dollar's worth for a nickel that day," he tells me, and then—realizing that it may not sound good to suggest that a nine-year-old was working over drunks—adds defensively, "Well, I'm not gonna say I didn't make a good buy, but I only had so much money."

After buying up what the drunks had on offer, Leonard and his father spent the weekend transporting the material to a local scrapyard, where they sold it for $36—nearly tripling their investment. "We never saw that kind of money in our life," he recalls. "And it became a very alluring business in the midst of the depression." In short order his father was dropping Leonard and his siblings at other dumps, buying up scrap, and selling it. It wasn't an overly lucrative business—the family lacked indoor plumbing, for example—but Leonard recalls that they were lucky, the kings of the heap on "poverty road."

By 1938 Fritz was no longer grubbing in city dumps with hobos. Rather, he was laboring in steel mill dumps on his own or with a couple of coworkers, in the employ of his father. So rather than spending his days in search of bottles, cans, and bones, he now spent them in pursuit of the far more plentiful leftovers of the steelmaking process. It was a savvy move. For the aspiring scrap man, the steel mill dump—and not the city dump—is where you build a business. Then as now, manufacturers and other large-scale businesses throw off more scrap than homeowners. And unlike homeowner scrap, it's typically sorted into grades and types of metal before you pick it up. There's no need to sit on the basement stairs, breaking the proverbial iron from the proverbial brass, as my grandmother and her siblings did in advance of brother Mort's bar mitzvah. Even better, whereas a house might throw off one pound of aluminum per week, a factory might do that in a minute. So rather than go from home to home in search of a thousand pounds of aluminum worth selling to a remelter, the smart small-time scrap man is always on the lookout for a factory that can give him a thousand pounds in a single load, even if he has to pay for it.

As Leonard was soon to learn, however, an aspiring scrap man's ability to make a living in a dump, or anywhere else, is directly proportional to whether or not anyone wants to buy what's been dug out of that dump (proverbial or real). Back when Leonard Fritz was starting out, steel mills had dumps where trash and the residues of steelmaking were thrown away. This latter category was represented by sand, bricks, and bits of steel that fell away during the manufacturing process. Of these steel bits, the most plentiful were the thin flakes that form on the surface of hot steel as it cools.

"Mill scale," as those flakes were called, was pretty much useless in 1938. It couldn't be melted into new steel, or made into anything else, really. That's changed over the years: today it's mixed into concrete and used in the manufacture of alloys as well as the production of new steel. But back in 1938 the technology didn't exist to do any of that, so—lacking an application—steel mills tossed it into the landfill. Fortunately for Leonard Fritz, though, some scrapyards were buying it cheap and then mixing it into other scrap. It was a weak, not particularly lucrative market, but it wasn't competitive, and a hardworking one- or two-man operation like Leonard's could be confident of a small profit without having to worry about another grubber undercutting them.

Nevertheless, providing mill scale to scrapyards was hard physical labor, accomplished with little more than a shovel and some screens. According to Leonard, the process was simple: take a shovelful of the mix of mill scale, dirt, and brick, toss it onto a screen, and shake it. What's left over is—in theory—semivaluable mill scale. It's the sort of work that makes a man out of a boy, and Leonard, approaching ninety, looks back upon that time wistfully. "When I was fifteen I didn't look like I do sitting here now. I weighed one hundred eighty-five pounds, I was five foot eight and had a twenty-nine-inch waist. Just all bull."

I look away and notice that the shelves to Leonard's right are almost empty except for a few knickknacks and a small painting of Jesus with the message: "I am the way." His desk isn't unlike those shelves: empty but for a phone, a candy bowl, and a small lamp with a green glass shade. It occurs to me that if he could, he'd rather conduct this interview out at a dump, and not here. Statements of faith aside, this empty space seems to have little to do with Leonard Fritz.

On November 24, 1917, *Scientific American* published its supplement 2186, which included a short article that began on page 328, "Junk Is America's Richest War Bride: The Result of Amazing Wastefulness." Published in the midst of American involvement in World War I, and of wartime shortages in key raw materials, the contemporary, educated readers of the story were no doubt surprised to learn that, out of sight, the sweaty mix of outsiders who picked through their trash not only were getting rich but had a trade association that held annual dinners at the Waldorf Astoria: "Before the war, the gross business transacted by the junk dealers of America probably averaged $100,000,000. Recently at the convention of waste material dealers held in New York City, it was announced that the combined business of the junk dealers of America was now more than $1,000,000,000 per year."

The scale of the scrap-metal business, and its profitability, always comes as a surprise, if not an affront, to those who aren't part of it. In part, no doubt this has to do with the popular image of the business as one populated by gangsters, bums, and thieves. It's often a class-based bias (in early-twentieth-century America, it also smacked of anti-Semitism), but to be sure, the poorer classes are where scrap entrepreneurs are born. Predictably, James Anderson, the author of the *Scientific American* story, aligned squarely with the classes that don't start junkyards: "The waste paper kings, who are making fortunes from the industry, do not have skyscrapers along Broadway, and you won't find any of them in any of the buildings around Wall Street. If you are looking for the factory of a waste paper king, go down along the waterfront and pick out the worst ramshackle building to be found down there."

At age fifteen, Leonard Fritz fell out with his father and left home with $3 and nowhere to go. All day, that first day, he wandered down the railroad tracks until he came upon Saddler's coal yard in Dearborn. Leonard needed a job, he recalls, but Al Saddler didn't have one to give. Still, Saddler saw something in Fritz, and offered to sell him his two-year-old Chevy truck for $300, payable at the rate of $1 per day, starting just as soon as Leonard got himself started in the scrap business. "And I just couldn't believe it," Leonard recalls. "My mind was all on different places I could start scrapping and things." Now, instead of having to pay someone to transport his day's work to a scrapyard, Leonard could do it himself. And he could transport—and sell—much more.

The job of the scrap man, as Leonard saw it, was to extract the value out of what everybody else saw as worthless, or couldn't be bothered to extract on their own. "One of my first jobs was getting the re-rod [steel reinforcing beams] out of a smokestack on Junction Avenue that had keeled over, you know," he tells me. "And I would beat the hell out of that great big stack and get that re-rod out. Big old sledge hammer." He was making between four and six dollars per day in those days, "which wasn't bad."

Elsewhere in Detroit, Armco Steel, a major supplier to the automobile industry, was testing a new type of steel furnace. But Armco was dealing with an expensive problem: preparing ore for the furnace cost roughly $100 per ton. So the designer of the new furnace, a metallurgist who happened to be the daughter of the mill's president, did some calculations and figured out that if you added a substance with the chemical makeup of mill scale, the preparation process would speed up and the cost would drop by $99 per ton, to $1. But where on earth could Armco get enough mill scale to feed its furnace? For reasons lost to history, the job of sourcing that mill scale landed in the hands of the metallurgist's young husband, who was visiting the Kelsey-Hayes wheel forging plant near Detroit on July 2, 1938.

As it happened, Leonard Fritz was busy screening a three-hundred-ton pile of mill scale in the parking lot of the Kelsey-Hayes forging plant that day, working at an agreed-upon rate of $1.25 per ton. Midafternoon, he looked up from that pile of mill scale and saw the husband of Armco's metallurgist walking toward him: "He had a homburg on and a real fine camel-haired top coat. And I had no shirt, you know, because I'm shoveling and working in the yard. He was a Harvard graduate. He didn't know the streets. And he said, 'Ah, how much of that mill scale do you have there, young man?'"

"I figure there's about three hundred tons."

"That won't be enough."

"Well, how much do you need?"

"What I'm looking for is three thousand tons."

Leonard Fritz was only fifteen, but he already knew a few things, one of which was this: there was a dump in Detroit where mills had been dumping scale for years. "You had to dig it up and screen it," he explains. "But it was all in one section." If the man in the homburg were really willing to contract him for three thousand tons, he'd need help digging it up, but even at his current rate of $1.25 per ton he'd have a guaranteed income stream of $3,750 coming to him, and that would be more than enough to hire some quality help. "I think I can scrape it up," he replied. "What would you pay for it?"

The metallurgist's young husband hesitated for a moment. "Probably thirty-two dollars per ton."

Leonard Fritz understandably lost it. "WHAT?" he exclaimed.

Looking back nearly seventy years from the vantage point of his corporate headquarters, Leonard laughs at what the metallurgist's husband did next. "He took the 'WHAT?' as 'You've gotta be shitting.'"

The metallurgist's husband, clearly unaware that he was bidding against someone who until moments earlier had believed that mill scale was worth $1.25 per ton, raised what he must have seen as a lowball price. "Well, thirty-six."

"And I said, 'I'll do 'er.'"

The market for mill scale had just been reset. Starting July 2, 1938, it was determined by how much money the steel mills could save by using it in their steelmaking process; it had gone from borderline trash to a crucial raw material. Leonard Fritz not only knew where to get it, he was experienced in how to get it. Those two words—*how* and *where*—are what thrust men like Leonard out of the ranks of peddlers and into the rarer company of men who accumulate large volumes of recyclable material, and considerable fortunes.

Still, it was a big job, especially for a fifteen-year-old, and it required Leonard to hire a couple of friends and buy some additional equipment, including two dump trucks, for $3,600 of his advance on the work. But the debts were of no concern; a few months later, shortly after his sixteenth birthday, Armco Steel presented the teenage Leonard Fritz with a check for $186,000.

It was still 1938.

In the early summer of 1941, the eve of America's entrance into World War II, the city of New York undertook one of the nation's first "scrap drives." The goal was to procure aluminum, a lightweight metal that, among other things, could be used in airplane manufacture. Old pots and pans were a prime target, but so were window frames, utensils, and even child's toys. Once collected, the aluminum scrap would be delivered to aluminum companies for remelting.

Traditionally, the junk man was the middleman, collecting scrap metal from households and delivering it to the smelters. Small-scale junk men knew where to buy old pots and pans, what to pay for them, and—most important—how to prepare them for delivery to the smelters, who would then melt them. They had the patience, experience, and profit motive to separate them by type and strip away any nonaluminum parts, such as the steel screws holding handles to pots, that might contaminate an aluminum furnace. It's the sort of work my grandmother and her siblings did in advance of bar mitzvahs.

In ordinary times, New Yorkers put up with the junk man who made his living from their old things. But wartime changed this attitude. Foreigners were suspect in America, especially the scrap dealers who over the previous three decades had made a good business out of exporting scrap metal all over the world, including—notoriously—to what had become known as the Axis powers. So well-intentioned New Yorkers, including New York mayor Fiorello LaGuardia, were understandably less than enthusiastic about turning over their pots and pans to the junk industry. Driven by suspicion, LaGuardia charged amateur community committees with collecting the junk and delivering it to the aluminum smelters instead. Historian Susan Strasser, in her seminal social history of America's junk industry, *Waste and Want*, offers the best, most concise account of the resulting debacle: "[The smelters] normally bought aluminum scrap that the junk men had sorted, while the drives brought them whole refrigerators and baby carriages, items containing as little as two ounces of aluminum in fifty pounds of bulk." In other words, with no financial incentive to sort the two ounces of aluminum out of the baby carriages, the community committees instead delivered piles and piles of good intentions—"I'll give my refrigerator to the war effort!"—that strongly resembled worthless trash salted with value (the aluminum). No surprise that piles of aluminum-containing scrap accumulated at smelters, unused. The intended beneficiaries of this mostly worthless largesse, the smelters, had no choice but to hire crews to clean up the aluminum—that is, to go grubbing in it, sorting out the worthless materials from the valued metal. Had New York left well enough alone, the job could have been done more efficiently, and probably more cheaply, by junk men.

Undoubtedly, World War II–era home scrappers felt good when they dropped their trash into the collection bins for others to sort out. In fact, they probably felt just as good as today's home recyclers feel when they drop their iPhone boxes into blue bins and set them on curbsides. But neither act really does much to help the actual recyclers—the companies that actually turn that material into new goods. And that's why, following the disastrous 1941 aluminum drive, the federal government turned to the traditional junk traders and scrap men to handle the collection of recyclable materials for the duration of the war, more or less.

During the first half of the twentieth century the scrap industry—dealing in rags, paper, metals, bones, and other commodities—grew rapidly. According to Carl Zimring, the scrap industry in Detroit, Leonard Fritz's hometown, expanded like a balloon: there were 60 firms in 1890, 127 in 1910, and 296 in 1920. By 1948 the United States was home to 3,044 scrap iron and steel firms, alone, accounting for almost $1.7 billion in sales, according to that year's Census of Business, prepared by the U.S. government. But that was nothing compared to what would happen when post–World War II American prosperity created the most affluent consumer class in history, with all the trash that would be associated with it. The American scrap business, already a century old, was just getting started.

The $186,000 that Leonard Fritz earned from selling mill scale to Armco Steel was spent, in large part, on the sorts of goods that would later turn into valuable scrap, transforming the American scrap business into the $30-billion-per-year (at least) business that it is today. After paying his younger brother, Ray, and two kids he knew from school, Fritz went out and bought three Ford Model A cars, loaded. "And of course the cops were all stopping us and warning me, and they knew we didn't have a pot to pee in," he says with a gravelly laugh. "They were stopping these kids because they were figuring we were in on something crooked. It wasn't going over very big. I was driving a brand-new Lincoln." For his mother, Leonard bought a house and a Pontiac convertible. What remained—and that was quite a bit—he kept in cigar boxes: "I didn't trust no bank."

But he did trust his business sense. The mill scale money allowed him to buy trucks and equipment, hire employees, and bid for bigger piles of scrap. Good weeks, he earned $1,800; bad weeks, the earnings dropped to "seven hundred, eight hundred bucks." It was a comfortable life, but when World War II broke out, he didn't hesitate to enlist. "I had this heroic feeling about me, you know," he says and then pauses. For the only time in a two-hour interview, he can't find the words to explain precisely why he did something. "The excitement of the—you see, my body's all gone, the frame's all gone, but the mind still has that excitement in it." He pauses again, longer this time. "There's gotta be something better here, something different we can do."

Before departing for Europe, Leonard left his business in the hands of family. Two weeks later, they sold it for $5,000. Looking back, he says that the equipment alone was worth $25,000. But there was nothing to be done. For the duration of the war, he made $52 per week, part of which he sent home to support his mother. He returned in 1944 with little more than a bad case of kidney stones and optimism. "I had the most uncanny luck that any man has ever seen," he tells me with a faint smile.

That's for sure. The postwar American economy was beginning to boom, and if there was one man in Detroit who knew how to make money off all the waste it was throwing off, that was Leonard Fritz.

Leonard Fritz's Huron Valley Steel Corporation eventually hitched its fortunes to the greatest waste stream that the world has ever known: the one created by American automobiles. But that development was still a couple of decades away. In the meantime, Leonard spent his days profiting from what steel mills tossed in the dump. By the late 1950s Leonard employed roughly 127 men, he told me, digging up dumps "all around the country."

It was hard work, but it allowed a small-time entrepreneur like Leonard Fritz to compete against some of the world's biggest iron ore miners. After all, both businesses—the dump recovery specialists and the iron ore miners—served the same steel mills. Grubbing, however, is cheaper than mining, the sort of thing that a small-timer like Leonard could enter, and thus Leonard Fritz was able to start a raw material supply business. By the late 1950s, he was big enough to start making iron in his own furnaces.

CHAPTER 3

Honey, Barley

The first thing to do when you open up a small American scrapyard in the morning is unlock the safe and count the money. That is, if you didn't take it home with you. Back in the 1980s and 1990s, when I was active as a teenager and young college graduate at my family's scrapyard in Minneapolis, we chose to leave it in the safe for two reasons: first, driving around with three grand in the trunk of your car is never a good idea (though my father often did it anyway); and second, you never know when a customer might show up, first thing in the morning, with several thousand dollars' worth of scrap to sell. If we needed more cash—and we usually did—someone could always run to the bank when it opened at 9:00 A.M.

That's how I remember it, at least.

During the summer months, my younger sister Amy would join my father and my grandmother at the front desk, counting money. After college, I joined them, too. But for most years it was often just the two of them, my father and grandmother, at 6:30 A.M., counting money into the register. You could see them there, the short paunchy man with the circle of hair atop his head counting the big bills, and the hundred-pound whip with the ice blue eyes counting the small ones. Inevitably, though, the mother-son moment was interrupted by a ringing phone, and my father left his mother to finish the job on her own, while he took the call in his office directly behind the front one.

There were two features to that office: a tacky wall clock made from a slice of a giant tree stump, purchased at the Minnesota State Fair, and a very large window that looked out upon the front desk and its cash register. From there, my father could not only watch his mother and whoever she might be paying but also see the television screens that pointed at his metal warehouse filled with aluminum, copper, brass, and lead, the scales where his scrap was bought and sold, and the metal yard where people dropped off everything from old cars, to mainframe computers from the 1970s, to giant drill presses from the nineteenth century.

As he slipped into his ratty office chair, he'd glance at the television cameras and then hit the line-one button on the phone. "Scrap Metal Processors. How can I help you?" It could be anything: aluminum cans, baseball bats, copper mesh from a chemistry laboratory at the University of Minnesota, whole automobiles, half of a refrigerator, silver-plated wire, a load of bathroom scales. Nothing was surprising, and everything had a price. "Tuning forks?" he'd ask the person on the other end of the line. "Maybe fifteen cents a pound, but I'll need to see them. Ask for Mickey when you get here." He'd then hang up the phone, turn around, and part the blinds for a brief look out the window at the nearby Minneapolis skyline, and the cars belonging to his arriving employees.

Right around that time, the cash register drawer slammed shut, ready for business, and my grandmother retreated into the glorified broom closet that she called her office. It was notable for a handful of features: the microwave oven, the refrigerator where she kept her stash of kosher hot dogs, and the odd array of brass figurines and loosely defined antiques that she had stolen from my father's employees, who in turn had stolen them from the metal warehouse, hiding them in places that only she seemed to know. It was the one room in the office with the scrapyard smell known to scrap men (and grandmothers) all over the world: tangy like metal and thin like a wire. I've smelled it on four continents, from small towns in Thailand to warehouses on the edge of Chicago. Each time, each breath, reminds me of my grandmother's office, and the piles of metal just beyond it.

While she cooked the hot dogs, one for my father and one for herself (and one for me, if I could be bothered to show up so early), my father would leave his office, cross the hallway, open the metal warehouse door, turn on the lights, and raise the loading dock door. While the lights buzzed to icy life, he'd take a walk around, squinting in the darkness at his inventory.

If he had a moment, my father might run his fingers along the edge of a carton filled with brass shavings generated by a factory in St. Paul; then peek into the bottom of a carton filled with automobile radiators delivered by a suburban repair shop. Near the front, there were always boxes of brass "drippings"—literally, the brass that dripped on the floor of a factory during the casting process; aluminum clippings, the clean scraps that fell away when a machinist cut a widget from a piece of aluminum; boxes of copper tubing delivered by plumbers; boxes of water meters; boxes of fine, shiny copper wires, dropped off by defense manufacturers who'd completed their smart bomb orders; steel bins full of aluminum cans dropped off by neighbors; cartons full of old PCs dropped off by well-intentioned environmentalists; more cartons overflowing with brass bullet shell casings picked up from a local gun range favored by cops, gangbangers, and, in my experience, dentists; printer plates from the local printer responsible for our business cards and letterhead; and forks, knives, and spoons that my father had bid for, and won, from a major airline.

It was fairly typical, though small, as scrap warehouses go. Still, it was more than enough to pay for a house in the suburbs and two private college educations (mine and my sister's). To my young eyes, though, the most amazing thing about that warehouse wasn't the scrap or the wealth, necessarily, but how quickly it all turned over. What that warehouse contained on Monday was never what it contained on Friday. The supply of scrap, and the demand for scrap, just never seemed to end.

In 2013, the seven thousand or so businesses that constitute the U.S. scrap recycling industry were responsible for transforming 130 million metric tons of recyclable waste into raw materials that could be made into new stuff. That's 130 million tons of iron ore, copper ore, nickel, paper, plastic, and glass that didn't have to be dug out of the ground or cut out of a forest. It also exceeded, by an astounding 55 million tons, the volume of recycled municipal solid waste—that is, recyclables dumped into blue, green, and single-stream bins—generated by homes, government offices, and businesses during that same period.

What's the difference between what a scrapyard recycles, and what gets transported to facilities like the Houston Material Recovery Facility? There's some overlap, but in general the scrap business handles everything that's not generated in the daily course of life in an office or home. Your old automobile ends up in a scrapyard; so do the metal grindings that fall away when an automobile manufacturer makes a new engine; the old electric meter on the back of your house ends up in a scrapyard (if the power company has the good sense to sell it); so do the power and telephone lines that connect to your house when they're replaced; the cardboard packing boxes behind your local supermarket go to a paper scrapyard; so do the unsold newspapers in newspaper boxes.

Altogether, according to ISRI, the American scrap recycling industry, a set of companies that buy, pack, and process everything from metal to rubber, employed 138,000 people in 2013. But for all of the traceable businesses, with traceable employees, there are just as many untraceable ones: everything from the organized gangs of scrap thieves who roam Detroit to panhandlers who stick their hands into subway waste bins in search of a Coke can. It's hard, I know, to think of a panhandler as part of any industry, but believe me, if the panhandler didn't pull that bottle from the subway bin, nobody else would. He's the bottom rung of the chain that moves up through your home recycling bin (he might steal the contents to sell rather than allow you to give them away) through my father, a processor and packer, to the companies that melt and transform scrap into new metal, paper, and plastic.

On weekday mornings at my father's scrapyard, the customers who showed up before 7:00 A.M. were strongly represented by plumbers, electricians, and contractors with scrap that they'd acquired on recent jobs—usually plumbing, wire, siding, and window frames. They weren't panhandlers, but they too were at the bottom rungs of the American recycling chain, collecting what a large company—like Waste Management—would never have bothered to collect because the volumes were just too small to be worth the trouble. Sometimes they brought a mere day's worth of stuff, just enough to be cashed in for a couple of cases of beer; and sometimes it was enough to pay for a first-class barbecue to go with that beer. Usually, though, the transaction fell somewhere in between those extremes. A handyman, for example, might arrive with several white plastic buckets. One might be filled with copper tubes used for bathroom plumbing; another might have old brass plumbing fixtures and perhaps a few brass electrical connectors; and the last might be filled with a lightweight mix of wires and an electrical meter or two.

My father, a man with a talent for gabbing with the random characters who appear on scrapyard docks, would saunter over with a gambler's confidence. "What've we got?" Then, without waiting for an answer, he'd pick up one of the buckets and place it on the kitchen-table-sized metal scale built into the concrete floor. To the surprise of many first-time customers, this was no digital scale, however, but rather a balance scale hung with weights that slid up and down a long arm. In the opinion of my father and most of the scrap world, such a scale is much more accurate than one outfitted with a mere digital output. Whether that is true, I still don't know.

As the electrician looked on, my father would slide the scale balances down the beam, achieve a very quick weight, and write the weight on an invoice. Then it was time for the second bucket, and my father would reach for it, quite often before that customer could get around to asking what the day's price on copper tubes was. "Ah, so what're you paying for copper these days?" the electrician might ask as he watched the second bucket slide onto the scale.

This always took some reckoning. Scrap metal, despite its association with trash, is as much a commodity as bushels of corn, barrels of oil, and ingots of gold. If a customer brought my father a solid copper ingot, the price would be easy to determine. Pre-Internet, my father would just go to the *Wall Street Journal*, look for the price of copper on the London Metal Exchange (LME) or the metal trading division of the New York Mercantile Exchange (COMEX), and offer that, minus a few cents to allow himself a profit. Then he'd sell it to a company that melts copper—perhaps the nearby foundry that cast copper into pots and pans.

A pound of wires isn't quite so simple to buy. After all, a pound of wire isn't a pound of metal; it's a pound of metal (often more than one kind) and insulation. The insulation doesn't weigh much, but you'll have to pay somebody to remove it, and separating different kinds of metals costs even more. So the price paid for the wire needs to reflect those costs, or the transaction isn't going to be profitable for the buyer. The experienced scrap buyers—people like my father—know by experience, if not instinct, what the metal recovery will be from a certain kind of cable. If they don't, they'll call someone who does. And that reflects perhaps the most important piece of accepted wisdom in the scrap business: you make your money when you buy scrap, not when you sell it. For example, if you buy scrap wire believing that it's 20 percent copper, and it turns out to be 10 percent copper, it's going to take a serious and highly unlikely upward swing in the London Metal Exchange before you can recover your loss. It's a rule that holds for small-time peddlers just as much as it does for large multinational companies.

In any event, once a scrap buyer knows, or thinks he knows, the percentage of metal in a load of scrap, he then formulates a price by looking up the LME or COMEX price for, say, copper, and subtracting the cost of processing from it. For example, wire collected by an electrician might be bought for only 20 percent of the COMEX price, and sold—if very lucky—at 40 percent. So, when speaking to other scrap dealers, or companies with some savvy in the scrap markets, my father wouldn't talk actual prices ("We're paying a dollar twenty-five"), but rather would quote in formulas based off the exchanges ("We're paying minus five").

Meanwhile, the electrician would pay as much attention to how my father manipulated those balances as he did to the prices quoted. In the small-scale scrap business, the former was often as important as the latter. For example, my father and I were both familiar with one Minneapolis scrap dealer who, when he weighed metal, placed his always ample cigar (often weighed down with a BB embedded in the tobacco) on one of the scale balances as he jovially chatted with the customers. Those extra few ounces of tobacco—and the BB—placed on just the right spot on the beam, seemingly innocently, shifted the weight in the scrap buyer's favor by several pounds. Other dealers, also in the Minneapolis area, took cruder approaches: I was personally familiar with one small-time scrap dealer who handed out pens decorated with buxom models in scratch-off bikinis (I was also familiar with the mother of this dealer, who purchased those pens on his behalf). As the distracted customers scratched off the bikinis, the scrap dealer swung the scale balances back and forth in a blinding pantomime of what it might look like if a cartoon character weighed scrap metal. Only in this case, it was just a shell game whereby the customer—dizzy with scratch-off bikinis—didn't notice how much weight he'd just lost on the scale.

But it isn't just the buyers who play games with the scale. Arguably,

it's the sellers who show true rapaciousness. Indeed, it's the rare scrapyard that hasn't at some point bought rocks wedged into aluminum
beverage cans, boulders packed into the trunk of a car, or scrap radiators filled with sand. And that's just the small-scale fraud. I once watched
as a Chinese paper mill unpacked bales of imported American newspapers, each stuffed with a cinder block to weigh it down, bought from a
well-known, very large American paper scrap company. Scrap importers worldwide are happy to share similar stories.

By the time my father had finished writing out the morning's first
scale ticket, his employees would have begun to arrive. One of those
employees might pick up the handyman's buckets and, depending on
what they contained, either dump them into a box with similar materials or push them off to the side until they could be paired with a bigger
box, guaranteed to arrive at some point in the next day or so. Meanwhile, one of our delivery trucks might be arriving, loaded down with
several washing-machine-sized boxes of copper shavings, generated
during the overnight shift of a factory across town and picked up twenty
minutes ago. That load would be worth more than all of the scrap collected by the plumbers and electricians who wandered in during the
course of the week (and there would be tens more such loads from various factories in the course of the week). My father, meanwhile, would
take the contractor's ticket into the office, drop it off with his mother,
and then retreat into his office, where more likely than not there was a
hot dog and a kosher dill waiting on a paper plate next to a list of what
he had in inventory, available to sell.

The customers for my father's scrap metal were divided into three groups,
with some overlap: mills, refineries, and foundries who bought scrap to
melt it into new metal; bigger scrap companies with the ability to buy
large volumes of scrap from smaller yards and then sell it for a high
price to a mill desperate for lots of scrap; and brokers who dealt with
both groups. No matter the customer, however, my father would always
start his morning phone conversations with them in the same way. He'd
greet the buyer, likely someone he'd known for years; they'd ask about
each other's families, talk sports, tell a dirty joke or two, and then they'd
get down to business—business unlikely to be understood by anyone
outside the global scrap industry. "What're you guys paying for Honey?

Um-hm. How about Barley? That's it? Okay, I'm sitting on a bunch of Ocean that I wanna move. Okay. And Birch/Cliff?"

Honey. Barley. Ocean . . . Birch/Cliff?

It's the secret, invented language of the global scrap trader, a recycler's Esperanto, if you will, with a lineage that goes back to the second decade of the twentieth century, at the latest. Back then, recyclers of old rags and clothing had a problem: How do a buyer and seller agree on a deal for a ton of old cotton rags, if every ton of cotton rags is different? The answer, first formulated by the National Association of Waste Material Dealers (NAWMD) in 1914, was to create binding specifications that said precisely what those rags should look like. If the rags, when delivered to the buyer, differed from that specification, the buyer would have grounds for a complaint, arbitration, or lawsuit. It was a concept that worked, and by 1917 the original three grades of rags (mostly used for papermaking) had expanded to twenty-three grades, including:

Extra No. 1 White Cottons. Large white clean cottons, free of knits, ganzees, canvas, lace curtains, stringy or mussy rags.
No. 2 Whites. Soiled white cottons, free of dump, street rags, scorched, painted, oily rags.
Black Cotton Stockings. To contain only black cotton stocking. White feet or edgings are permitted.

Snicker, if you will: but for a paper mill looking to control the color of its writing paper, a specification that guarantees black socks, and no other color, is serious business. By 1919 scrap specifications had moved past socks and paper into metals, and by the early 1950s they were well-established instruments of the international scrap trade.

There was, however, one problem: the fastest means of consummating business deals in those days was the Teletype, and the Teletype companies charged by the character. So to simplify communication and cut down on expensive Teletype bills, scrap dealers agreed on a set of four- to six-letter words to represent the various grades of scrap recyclables that they traded. For example, *Talk* became shorthand for "aluminum copper radiators," *Lake*, for "Brass arms and rifle shells, clean fired," and *Taboo*, for "mixed low copper aluminum clippings and solids."

Thus my father, when speaking of clean wire, used Barley, the specification devised by ISRI, the Washington, DC–based recycling trade association directly descended from the NAWMD. And that specification was, and is:

> **Barley. No. 1 Copper Wire.** Shall consist of No. 1 bare, uncoated,
> unalloyed copper wire, not smaller than No. 16 B & S wire
> gauge. Green copper wire and hydraulically compacted
> material to be subject to agreement between buyer and seller.

If in the course of his morning conversations my father agreed to sell a load of Barley, then he would be contractually bound to—among other things—make sure that he didn't deliver insulated wire, such as Christmas tree lights. The specification, after all, specifies *bare* wire, and delivering anything but bare wire would be a breach of contract. As a result my father and his employees expended considerable effort in sorting the often random tangles of wire and cable that they received in the warehouse, pulling out bits of insulated wire from the bare wire to make sure that the specification was met. If it wasn't, the buyer had the right to reject the load. More often than not, though, a buyer with a "claim" against a load of wire would just negotiate a lower price, costing my father a part of his profit margin.

Specifications aren't important only to scrap-metal dealers. The differences between various grades of paper remain just as significant as they did a century ago, and serious resources are devoted to identifying and separating them. In 2006 I watched as dozens of women spent their day ripping cardboard notebook covers from scrap school notebooks at the Rama Paper Mills roughly sixty miles outside Delhi, India. The reason was simple: cardboard is more expensive than white notebook paper, and used in different ways at a paper mill. The exporters of the notebooks, located in Dubai, couldn't afford to do the sorting profitably—Dubai's labor costs are far too high—so they sold them to the Indian mill at a discount reflecting the fact that many grades of paper were mixed in the jumble. For Rama Paper Mills, the "upgrade" from cheap, recyclable notebooks to expensive cardboard and less expensive white paper was highly profitable.

Today, there are hundreds of individual scrap specifications. Some are specific to certain countries (Korea and Japan have their own specifications), but the most dominant and widely used ones are the North

American ISRI specifications. They are not static: they evolve with the nature of what people throw away, and with the technologies used to process those throwaways. Notably, the committee that writes them has a sense of humor: in 2007, long after the demise of the Teletype, they decided that Tata, Toto, and Tutu could serve as handy shorthand for three types of aluminum scrap. These, along with the older, more widely traded specifications, are as much terms of the trade as *dollar, ton,* and *freight.* It's a phenomenon that I was reminded of when, in 2010, I was in Jamnagar, a small city in northwest India whose economy is fueled by thousands of brassworks powered entirely by scrap metal. One afternoon, while walking through an industrial park with a large-scale scrap importer, we passed a scrawny scrap peddler with little more capital investment than a bicycle. At the sight of me, a rare foreigner in those parts and a likely source of imported scrap, he pointed at the load of mixed brass scrap in the trailer behind his bike, smiled, and said: "Honey."

I knew what he meant, and so did the person who exported that Honey to India. More important yet, the person who would buy it from him surely knew the definition as well. These odd terms aren't just relics of an eccentric business; they're the means by which random waste is transformed from a jumble of trash into a product that can be sold. After all, if you can't describe what you're selling, you likely won't be able to sell it. Specifications drive recyclables to the people who want them. Sometimes those people are in China, sometimes in India, and—in a very big way—sometimes they live and work in places like the American Midwest.

Just after seven on a weekday morning in mid-August 2011, Guy Dumato, native of Fort Wayne, Indiana, is driving a very large, very expensive black pickup truck down the streets of his hometown. I'm seated beside him, separated by what feels like yards in the massive truck cab, watching as he drinks from an oversize plastic mug of coffee. He's a compact and muscular figure in his late thirties, wide awake with the ready-to-work caffeine itch of a man long accustomed to early shifts.

Guy is a manager at OmniSource, one of the world's biggest scrap companies, which in 2007 was purchased for $1.1 billion by a publicly held steel company. Years ago, when I used to work in my father's business, we sold scrap to OmniSource. But until now, I'd never seen what happened to that scrap. As we drive, he tells me he started out modestly with OmniSource, as little more than a laborer, doing whatever required a strong back and a willingness to work. Curiously, though, Guy doesn't remember the past in terms of the jobs that he used to do. Rather, he remembers it in terms of the price of scrap metal.

"I remember when copper was sixty cents a pound," he recalls. "And a trailer full of it cost twenty-four grand." Those were the inexpensive old days, mostly ended in the 1990s, before India, China, and other developing nations began to demand raw materials—among them scrap metals—to build infrastructure and lifestyles. On the morning that I was traveling with Guy, the U.S. economy was still in recession, but demand from the developing world, such as China, where cities sprout suddenly from farm fields, had driven copper to over $3.50 per pound; the $24,000 trailers of scrap copper that Guy used to load were now worth closer to $150,000. That's the difference between a Ford Focus and a Ferrari.

"This is it," he says to me, and we turn right, into a small parking lot beside a tall, brick warehouse building on a quiet street on the edge of a residential area. It doesn't look like the kind of place that would house the world's biggest anything. But it does: this is the world's biggest factory for chopping up—that is, recycling—cable and wire into its various components, mostly copper, aluminum, and insulation.

Guy leads me into the warehouse. Not far from the door I see a colorful tangled pile of cables, some as thick as softballs. The sources are as mixed as the colors: some might be copper telephone cables recently replaced by fiber optics; some might have been dug up by a utility, replaced for an upgrade required by a wind power project; and some of the odds and ends might be manufacturing scrap purchased from one of the last survivors of the once-mighty American wire manufacturing industry. But most of the material, roughly 60 percent, is purchased from other, smaller scrapyards—such as the one I grew up around—throughout the United States, who in turn bought it from customers ranging from handymen to factories.

No statistics exist on the volume of waste cable and wire generated every year in the United States (or anywhere else, for that matter), but OmniSource's Fort Wayne chopping plant offers a rough perspective on just how much there *could* be: during a twenty-four-hour shift the plant can process 450,000 pounds of wire and cable—roughly the weight of the Statue of Liberty.

OmniSource's isn't the only North American wire-chopping plant: there are at least forty more, though none come close to OmniSource's size or sophistication. And that's just North America: scrap companies in India, China, and Vietnam, just as hungry for scrap as OmniSource, operate thousands of small-scale wire processing plants where hand labor, and simple machines substitute for the behemoth I've been invited to see. But first Guy takes me around a corner. "You've gotta see how we know what we're doing."

I follow him back into a compact, windowless room where two wiry young men, one in a white T-shirt that shows off his muscular arms and tattoos, use pincers to carefully tease apart two-inch samples of cable. Guy draws my attention to a twenty-foot-long pegboard wall covered, floor to ceiling, with hooks that gently cradle thousands of two-inch slices of wire and cable. To the left, the mostly thin wires form a tight black pinstripe; in the middle, things become thinner and more colorful, augmented here and there by large cross-section slices of tight, shiny copper cable, perhaps four and even five inches across, that remind me of nothing so much as grapefruits waiting to be sprinkled with sugar.

It's a hypnotic display, an unexpected work of accidental art that also happens to be a fairly comprehensive inventory of all the ways that Americans have transmitted power and information over the last thirty years or so. But the REFERENCE ONLY sign that hangs above it suggests a much deeper meaning. For OmniSource, which recycles more wire in one place than perhaps anybody else in the world, this wall is one means of ensuring that they know exactly what they are buying. Each piece of wire, each segment of cable, is wrapped in a white label that describes its contents precisely. Guy reaches for a random segment—a figure-eight-like piece made up of two strands, one larger than the other, held together by black insulation—and shows me the label:

FIGURE 8: $3\frac{1}{8}$"

37.81 #1 CU

8.82 CU FOIL

21.26 FE

The translation isn't hard. Figure 8 is a type of wire hung above the ground from utility poles, and this sample is of the 3 and ⅛ inches variety. Figure 8 wire is composed of two strands. One carries power or telecommunication signals—that's the 37.81 CU or, in plainer English, the 37.81 percent copper. The other strand is called the messenger cable, and its job is to physically support the other wire while it hangs from the ground—in this cable, 21.26 FE or, in lay terms, 21.26 percent steel. The other element marked here is copper foil (8.82 CU), typically wrapped around the copper wire to provide insulation from electrical and other types of interference. In scrap industry terms, that's a 46.63 percent copper "recovery." The higher the recovery, the more valuable the wire.

Across from the wire and cable reference wall, the two young men continue to pull apart cable samples. As they strip away insulation, copper threads drop into tea-saucer-sized stainless steel pans, which are then placed on digital scales and weighed. It looks like a dissection—which, in fact, it is: it's a precise dissection of the many millions of pounds of wire that OmniSource buys every year. These young men and their wire dissections provide real-time data on what OmniSource is buying from its hundreds of customers—a necessary supplement to the reference wall. "The thing is, we're still seeing new stuff, new kinds of wires and cable," explains Guy. "And new recoveries." He points at a random—to me—piece of cable on the wall. "We used to recover in the low sixties from this," he says, referring to the percentage of copper recovered from wire and cable. "Now we're recovering in the low fifties."

The reason for the downward trend is—paradoxically—the quintupling of global copper prices over the last twenty years. Cost-conscious manufacturers who once used lots of copper in their products, now use other, cheaper metals in its place (aluminum conducts electricity, too, but costs substantially less). "If we get a truckload of [mixed] wire, we'll probably sample it twenty times," Guy says with a shrug. "You have to if you want to know what you're buying. The scrap dealers who sell to us were all used to one recovery, and now we've gotta tell them there's another." Or, put differently: most of the old-line scrap dealers still aren't accustomed to a world where China, the world's largest copper consumer, dictates not only the price of copper but also the amount of copper used in a power line.

"You ready?" Guy asks.

I nod.

Guy gives me a hard hat, safety glasses, and earplugs, and we turn the corner.

The chopping line—or granulator, as it's sometimes known—is a multistory giant, extending several hundred feet into a warehouse bathed in natural light and surrounded by the roar of motors and the hiss and scream of metal fragments dropping onto metal. It's not just loud; it's heavy-metal-concert loud, even through the earplugs.

Twin conveyors at the front of the system are the only objects that seem familiar; farther out, I recognize hoses and more conveyors, and— way down the line—a copper escalator of sorts grabs handfuls of shiny copper fragments the size of peppercorns and lifts them into the system, where they disappear.

"Sound is really important in this process," Guy hollers as we walk to the front of the chopping line. Behind us, a loader rumbles into the room and dumps a tangled mass of cables weighing hundreds of pounds onto a shaking pan roughly the width of a bookshelf, and maybe fifteen feet long. As it shakes, the wire slowly moves forward, until it lands on a conveyor that sends it up and drops it into spinning blades. When the wire hits them, the chopper makes a deep, hollow groan that can be felt in the floor, and in my bones. "Ears are VERY important at this stage," Guy yells, and points me in the direction of a man completely covered in scarves, glasses, long sleeves, a respirator. He looks like a guerilla fighter. Guy points at the guerilla's right foot: it's operating a pedal that controls the rate at which the vibrating pan shakes, and thus dumps wire onto the conveyors. "Looking and listening for the right volume, listening for impurities," he calls out as the room groans with another tight knot of wire flowing into the machine. "The sound of that granulator makes a real difference. Tells you what's going through it. Real art to this, requires some experience."

I follow Guy down the line until we stop below the enclosed section of the chopper, where blades reduce the wire to roughly one-inch fragments. That's the easy part. The hard part, what most of the length of the system is devoted to, is separating the various kinds of metals from each other, and from the plastics. Some of that separation is commonsense: magnets are deployed to pull out steel. Some of it is a higher-tech version of Raymond Li's Christmas tree light recycling system in Shijiao: the chopped-up metal and plastic is run over a vibrating table beneath which air is blowing; the heavier metal will flow in one direction, the lighter plastic, in another.

Guy invites me to follow him onto the scaffolding that wraps around some of these tables, and I watch as fragments of clean metal shine like water falling off conveyors. At Guy's direction, I look below me, and there's the clean stream of metal, like a fast-flowing creek, eventually dropping into large heavy-duty plastic sacks capable of holding washing machines, which instead hold as much as four thousand pounds of metal. Around the scrap industry, they're known as "Super Sacks."

We head out the back loading dock, where I pick up the hiss of plastic falling down from a conveyor onto large piles of insulation accumulating in concrete bays. There's no lack of places to sell metal in North America, or worldwide. But the insulation that is the inevitable by-product of wire and cable recycling is more troublesome. Different types of plastics don't melt well together, and—so far, at least—nobody has developed technology to separate them profitably. Meanwhile, American manufacturers are far more quality-conscious than the plastic slipper sole manufacturers to whom Raymond Li sells Christmas tree light insulation. So, lacking customers that will buy large volumes of mixed plastic and rubber, OmniSource and other North American wire choppers generally send the insulation to landfills.

The quantity of insulation in a wire or cable plays an important role in whether OmniSource shreds it in Indiana or ships it overseas for processing. The more insulation, the more likely it is that OmniSource will export it, though there are other factors, including the type of metal packed into a cable. There's no hard-and-fast rule for the copper-insulation threshold, but it's the rare North American or European wire chopper that will take anything less than 60 percent metal. Christmas tree lights, by contrast, are 28 percent copper and brass, and thus they go overseas, where low processing costs, huge demand for copper of any kind, and ready markets for mixed insulation make them highly desirable.

By contrast, the giant sacks of metal at the end of the chopping line can contain copper of 99.9 percent purity. Think about that: when I visited in late summer 2011, copper was priced around $3.50 per pound—meaning that a 4,000-pound Super Sack of copper was worth roughly $14,000.

Still, Guy cautions me to not get too excited. "The margin on this stuff is pennies," he says, referring to the high cost of processing wire in the United States. "What makes it profitable is the fact that we do ten million pounds of it [per year]." He leads me around a corner and into another long warehouse where, laid out in neat rows, are hundreds of Super Sacks, each with a unique mix of chopped copper. Not all can be said to be 99 percent pure—most fall between 96 and 99 percent—but there is no doubting the millions of dollars idling in the sunshine coming through the high windows.

We stop next to a random sack, and Guy opens it to reveal quarter-inch fragments of copper that shine like gold. He jams his hand into it and lifts the metal into the sun. "This is 99.75 percent. It's not our best product." The 0.25 percent that makes it less than best is composed of the yellow fragments that I notice catching the light. They're brass—they might have once been electrical connectors attached to the ends of wires—and brass is difficult to separate from copper. Making matters even more difficult, brass is a mix of copper and zinc—so if you have brass in your copper, you really just have brass, not copper. Not all is lost, though: OmniSource knows brass mills that like to purchase copper contaminated with a precisely defined percentage of brass. More often than not, they remelt it into new brass. Some are in India, and most are in the United States. For them, this is prime raw material, manufactured to a precise specification. Nothing is wasted.

"There's a market for that?"

"A big one."

That market is the natural, remarkable end point to a winding supply chain that might start with a bucket of old wire sold at an Indiana scrapyard, or a USB cable dropped into a recycling bin in New York City. Along the way, wire is bought, sold, chopped, and sorted until it reaches a place—and a stage—where somebody can afford to make it into something new. The chain is commonplace: refrigerators, plastic bottles, and old textbooks follow the same path, the only difference being the processes used to turn the used-up goods into raw materials, and the locations of the people and companies who want to buy the results. Twenty-five years ago, most of those people and companies were in North America; today, they're everywhere.

I witnessed that change from my family's scrapyard and, later, traveling to scrapyards all over Asia and the world.

* * *

Back when I worked at the family scrapyard, my father would spend most mornings with a Minnesota manufacturer's directory, engaged in cold calls. However, he wasn't trying to sell anything to the random factories that he called. Rather, he was trying to buy scrap, and this—ironically—was what we called "sales." "Who currently handles your scrap metal?" he'd ask the new voice on the other end of the phone line. "Um-hm. Well, I bet I can beat their price."

On some occasions he could. But even if he did, that might not be enough to earn the right to buy a small factory's excess scrap metal. Rather, my father would also have to convince that scrap supplier—say, a manufacturer of food processing equipment with several hundred tons of aluminum grindings, annually—that he could pick up their scrap on time, provide quality service, and guarantee, perhaps, tickets to Minnesota Twins and Vikings games when needed. It was not lost on either of us, however, that other scrapyards offered similar perks—including better Vikings seats—to entice local factories into selling their metal. And Vikings tickets, frankly, are the least of it. Some scrap peddlers happily slip envelopes of cash to the dock manager at factories, with the understanding that he'll look elsewhere when they drive away with the company's barrels of valuable scrap metal; in China, lavish dinners, often concluded with prostitutes, are often just the base entry requirement if you want to even talk with certain factory bosses about their scrap.

Fact is, the competition for scrap remains ferocious in every market, worldwide. It's like food, really: if you don't have it, you die, and if you don't have enough, you don't grow. So you go out looking for scrap, cold-calling factories, utilities, and municipalities, offering to beat the price and service of the competition, hoping to take away their scrap—while they try to take away yours. Nine-year-old Leonard Fritz, grubbing in the dumps with hobos who tried to steal his scrap, experienced it young; my father, bidding against other Minneapolis scrap dealers, experienced it throughout his career.

The only hedge against the competition, really, is a big book of customers. I think, at the peak of my father's business, he was picking up scrap from roughly two hundred small manufacturers, utilities, and municipalities. Some of those customers were larger than others, but he could comfortably lose one or two and still stay in business. Bigger companies, including companies who competed against my father, might have hundreds of customers, and they could lose tens and not notice. But no matter the size of the company, small to multinational, they compete just as fiercely for the right to pay money for scrap. In that way, it's just the reverse of a normal business where you choose your suppliers (in fact, they compete to sell to you) and market to your buyers.

Thus, in the scrap industry, there's an axiom: it's hard to buy scrap, and easy to sell it.

I can't recall, precisely, when the first Chinese scrap buyer appeared at the front window of my father's scrapyard. It was probably around 1994, right around the time that China had begun to deregulate key industries, and private entrepreneurs had decided that scrap metal was the business where they'd strike it rich. It was a good bet: China was at the front end of a drive to become one of the world's great economies. It had labor and government support; the only thing it needed was raw materials. Digging mines was one way to obtain those raw materials; the other was to go to the United States, the place that many scrap traders call the Saudi Arabia of Scrap, the land where there's more scrap than the people can handle on their own. It's a funny nickname, Saudi Arabia of Scrap, but it's not meant as a compliment. Rather, it's an opportunity to exploit.

Those first Chinese traders are a blur to me (they weren't the first Asian buyers, though: we'd been selling smaller quantities to Taiwan for years). I just remember Chinese faces, broken English, and a willingness to buy everything in our inventory. "You have number-two wire?"

Sure, we have number-two wire. We also have customers for it. "How much do you want?"

"Can we see it?"

So we'd go out to the warehouse, and after a quick inspection, they'd ask to buy all of it. My father would offer a price—one significantly over what our customers in North America were paying—and they'd accept on the spot, no question. Then, if they had the time, they'd spend the remainder of the afternoon watching as our entire inventory of wire was loaded into overseas shipping containers and readied for shipment to a port in China that only they knew as an actual place. For me, and probably most other scrap men at that time, Foshan was no more real than Atlantis.

CHAPTER 4

The Intercontinental

Growing up in Minnesota, I was never interested in China. Like most midwesterners, I thought the continental United States that extended out in all directions was territory enough to explore. If I were inspired to leave the country, Canada and Mexico were the likely distant-enough destinations. No surprise, I went to college in midwestern Chicago, and when it came to study abroad, I went to Italy. That was far away, for sure, and I suppose it made me wonder whether there was more wandering to be done. But then I returned to the U.S. and didn't really think about leaving again. There were, after all, the western deserts to explore, and even today that's where I most like to spend my time.

So far as I know, my grandmother never left the country except to visit Canada. And as it happens, my father didn't travel much, either. What I do recall is that I had a second cousin, Chucky, in South Texas, and he used to move scrap back and forth over the border. My father visited him once or twice, and would offer frequent advice over the phone. Other than that, I remember my father flying to Ethiopia in the early 1980s to broker a scrap-metal deal that didn't work out.

In the mid-1990s, when the family scrap business started trading in earnest with China, the transactions were all but local. Chinese traders arrived at our door, paid cash, and left with the scrap. It was international trade, sure, but it was trade that we could do from home. In the midst of the 1990s boom, my father flew to China for a couple of quick trips, returning with photos of Chinese scrapyards filled with people and colorful piles of wire. But I always suspected that those trips were nothing more than excuses to travel, if not enjoy some big nights out on the (Chinese scrap) town. The only lessons learned, so far as I could tell, were that the Chinese were becoming rich, and that they'd be hungry for scrap metal for a long, long time. They weren't bad lessons—over the last two decades, giant fortunes in scrap metal have been made on those observations.

Ironically, though, my family didn't manage to make one of those fortunes.

Recently, I've heard it said that if you owned a scrapyard and didn't become rich over the last two decades, you were either dumb or incredibly unlucky. I always chuckle when I hear that, even though it stings. Not only did we not make a fortune; our business actually suffered and even shrank during the boom years. Then again, the fact that the family business remained in business at all through that period is an accomplishment in its own right; if the markets were just, we should have been bankrupted.

Credit, in all its forms, belongs to my father. He's a talent, a scrap man to the core, one of the great wheelers and dealers in an industry that sends them spinning off like tires down a hillside. But that was not enough to make him happy. I've met or heard of other scrap men who in midlife give up, or grow bored with making deals for the sake of making deals. Typically, like my father, they're above-average-intelligence men who grew up in the industry before it was a "green" business, before there was an ideology, a meaning, a purpose, to it. Sure, it had its nonmonetary challenges—notably the cold, hard creep of government regulation—but those were irritants more than enjoyable problems to be solved.

So what happens to a scrap man who grows tired of money for money's sake, who doesn't respect his own talent for looking at a pile of somebody else's junk and making a bundle from it? Sometimes they turn to women; sometimes they turn to booze; and sometimes they do things like set up cabanas selling tequila on Australian beaches and call it a career. As it happens, my father toyed with the last option but chose the first two.

For much of the 1990s, in fact, while I was associated with the business, he was heavily intoxicated with booze and other substances. Needless to say, a permanently inebriated CEO brings all kinds of problems, no matter the industry; but in the small-time scrap industry, where most transactions are cash and your employees are a constant threat to rob you blind, it's as good as throwing money into a wire chopper.

Back then I was a young, inexperienced honors graduate in philosophy, ambivalent about the idea of devoting myself to the scrap-metal industry. There were other things I wanted to do: write songs, write novels, get a Ph.D. in evolutionary biology, and fall in love with depressive women. But when your family—and your family business—is in trouble, you do what you can. So I made one of the best choices of my life and went to work closely with my grandmother at the scrapyard. We did our best to keep cash from disappearing, and we labored to send my father to some of the finest chemical dependency treatment centers in the United States. All the while, I became adept at shutting down bank accounts and lines of credit as we balanced whether we were better served by paying the banks, our employees, or my father's habits.

It's amazing that we stayed in business. I distinctly recall the afternoon that I had to go to the bank holding our line of credit and explain—humbly—that my father was in treatment, so please don't cut us off. I have an even clearer memory of needing to check my father out of a treatment center in Florida so that he could sign for another line of credit from a financial institution, giving the creditor the false impression (created by me) that my father was spending a month sailing on the Gulf Coast (I'm all but certain that he'd never once set foot on a sailboat). In any event, by my late twenties two things were becoming painfully obvious to me: first, my father was never going to embrace sobriety as my grandmother and I had hoped; and two, I had no future in a business where the top manager was extraordinarily talented but mostly absent.

It wasn't exactly a lost cause—my father had demonstrated a talent for doing just enough to keep the company afloat—but it wasn't a future. I needed a life, something beyond that scrapyard, even if "beyond" meant that I couldn't have lunch with my grandmother several times per week. But it wasn't just having lunch: there's nothing more enjoyable than sitting at a bank of video monitors with your grandmother and catching your employees stealing. We had so much fun.

Back then, I'd whittled myself down to two passions beyond scrap. The first was music, and the second was journalism. I tried my hand at music with little to no distinction, and gave it up at the point in one's twenties when these things are supposed to flame out. But journalism, for reasons I don't entirely understand, seemed to work. In Minneapolis

I started out freelancing for magazines, quickly working myself into bigger and bigger assignments. Then, after a year or two, I was given the opportunity to do a freelance assignment on scrap in China, and I took it without hesitation.

What a terrible idea.

First, I didn't know the language. Second, I'd never been to Asia. And third, the family business was still flailing in Minneapolis. But my grandmother, daughter of the scrap industry, encouraged me to go. "You have to live your life," she told me. "You've got to *do*." I don't think she expected that I'd remain in China for a decade; I sure didn't. If she did, I suspect that she would've told me what she told others: that she really didn't want to see me go. If I'd known how badly she wanted me to stay, I would've stayed; if I'd known that I would be gone for her last years, I would've stayed. But it was supposed to be only a brief trip, and then I'd come home and find a job, perhaps at the local paper.

I went to China with a mostly clean conscience and a handful of assignments. Of those, the one that intrigued me most was still the scrap-metal one. I'd seen my father's photos of China's scrapyards, but I didn't really believe them. I needed to see them on my own, to make it make sense to myself. All that knowledge I'd gained around the scrapyard over the years wasn't going to be wasted.

I remember the first time I reported in Foshan, China, population 7 million.

I flew into Guangzhou Airport, where I was met by a scrap dealer, his sleek BMW, and a fresh-from-the-countryside driver. It was 2002, and Foshan wasn't much more than a spread-out set of underdeveloped villages somewhere west of a Chinese wherever. I'd only been in-country a couple of weeks at that point, and I'd had trouble finding Foshan on a map. This all seemed like a bad idea.

The drive from the airport traversed newly built highways and not so newly built country roads lined with high-voltage power lines that sagged to a few feet off the ground. Overloaded delivery trucks were the dominant means of transportation, jamming up the roads and—when there were shoulders—the shoulders, too. Back then it took almost two hours to reach the faux-rococo Fontainebleau Hotel, a yellowed porcelain doily in the heart of Foshan's Nanhai District.

By then, Nanhai was already one of the world's biggest processors of scrap metal, and you only needed to walk into the lobby to know it. Set amid lush, manicured landscapes that would make Louis XIV blush, cigar-chomping scrap dealers from around the world sat in baroque chairs and discussed where they'd get a decent hamburger when they made it up to Shanghai on the weekend. But that wasn't all: at any hour of the day, you could walk into the lobby of that hotel and find at least a couple of Caucasian scrap exporters having tea, coffee, or whiskey with a couple of Chinese scrap importers while some of Guangdong Province's finest prostitutes sashayed by, on the way to visit clients upstairs. If you needed to know the price of insulated copper wire—well, the global market was being made right there, all day and all night long.

Jet lag defined much of what happened in the Fontainebleau in those days. I remember seeing scrap guys consuming breakfast at midnight, steaks at 7:30 A.M., and poorly mixed cocktails any time at all. But that was just as well, because scrap processing was (and often still is) a twenty-four-hour-a-day activity in southern China. It had to be: two decades into the country's modern development, everything was starting to accelerate: airports, highways, apartments, cars. And everything, needless to say, needs metal.

Take, for example, subways: on the day I moved to Shanghai, it had precisely three subway lines. Ten years later it's the world's largest system, with eleven lines and 270 miles of tracks. However, China lacks ready access to sufficient raw materials of its own to build all those subways, so in very short order it's become a net importer of scrap copper, aluminum, steel, and the other metals needed in the infrastructure of a modernizing society.

Back then, if you were jet-lagged and had an amenable scrap-metal host (and they were all amenable if it meant access to American scrap metal), you could head out to the scrapyards in the dead of night. You'd arrive in the processing zones via expensive cars that zigzagged down a narrow brick-lined alley, out into a boulevard with murky, poorly lit signs, back into an alley, finally pulling up at some metal gate indistinguishable from other metal gates. The driver would honk, the owner would roll down his window so the guard could see him, and a worker would push aside the gate. Then you'd drive into a wide lamplit space, the headlights bouncing off piles of metal fragments, giant bales of wire, and, off to the side, a shed where two or three men—it was mostly men—fed scrap cables into machines that ran an incision along the insulation. Nearby, another team—often female—used that incision to pull away the insulation and expose the copper wire.

What I saw was so alien—except for all of that scrap. I knew what that was. It looked like what we used to send to China, only now it was *in* China.

Meanwhile, over in the farthest corner of the yard, the flicker of flames might send black smoke into the not-quite-as-dark night. The smell would be noxious (and, depending on the wire, dioxin-laced), but the goal would be anything but: profit. Wires too small to run through the stripping machines were a favorite item to burn, but anything would do if copper demand was strong; in the morning, the copper could be swept out of the ashes. One night, I recall clearly, I saw a row of a half-dozen electrical transformers—the big cylinders that hang on power lines and regulate the power—smoking into the night. When I realized what they were, I backed off: older transformers contain highly toxic PCBs. But nobody seemed to mention that to the workers who, through the evening, poked at the flames. I didn't like it, but there's not much to be said when you're standing in the middle of a scrapyard in a village you've never heard of in a province you've just barely heard of, as the guest of somebody you've just met. I wasn't sure that I was in much position to be complaining, anyway: I'm a child of the industry too.

To be honest, I was shocked by the number of people who worked in these scrapyards, and by their low pay. But I was not shocked by the menial jobs, and I was not surprised by the pollution. After all, my grandmother and her siblings cleaned metal into adulthood, and her younger brother, Leonard, told me that he knew how to "break" a motor—that is, take it apart with hammers and pliers, and extract the copper—as well as anybody in the Twin Cities. That's what you do when you've got nothing else—and their generation didn't have much else.

That wasn't the only thing the Chinese and my family had in common. For example, I'm not ashamed to admit that my family often paid contractors to burn our wire in farm fields outside Minneapolis (we also ran an aluminum smelter with an open smokestack—arguably a worse offense). If it couldn't be burned, it would've been landfilled, and so we were doing what countless other scrapyards were doing in those days: using the cheapest means available to clean up other people's messes. Those days are over (for my family, at least) but I know of people who still do it in North Dakota—and there isn't an impoverished Chinese farmer among them.

To be sure, Foshan in the early 2000s was far more polluted than anything I saw in the United States while growing up in the 1980s and '90s, and surely more polluted than what my great-grandfather knew in his early years. But from my perspective, that difference was a matter of scale, concentration, and history. For better or worse, they weren't doing anything in 2002 that we didn't (or wouldn't) do in 1962. They were just doing much, much more of it. And as dirty as it might have looked at times, I didn't get the sense that the people around Foshan felt that scrap was "dumped" on them. Instead, they actively imported it, or they migrated from other provinces to work on it.

The pay, after all, couldn't be beat, especially if you were uneducated and illiterate. Depending on the scrapyard, salaries might be anywhere from 10 to 20 percent higher than what the local high-tech factory might pay. By U.S. standards, though, it wasn't much: maybe $100 per month plus room and board. Still, if your prospects were limited to a life of subsistence farming, that was more than enough money to send home to pay school fees. The next generation would have a better life, and the negative health consequences of scrapyard conditions could be worried about later.

In 2011 I fly into Guangzhou on one of my twice-yearly trips to its scrapyards, and lo, there's a subway that will take me to Foshan in less than an hour. Nanhai, which had once felt to me like a Wild West outpost divorced from all non-scrap-metal reality, is now another suburb of yet another Chinese megalopolis (Guangzhou: population 20 million plus). As I climb out of the station, I glance around me: I'm at the intersection of two busy, newly paved roads and four pieces of entirely empty farmland. Two blocks away, however, is the incoming wave of wealth: dozens of construction cranes hovering over dozens of high-rises, some as tall as thirty stories, each taking a bite out of open space recently home to farms. I roll my suitcase in their direction, through crabgrass and dirt littered with paper instant noodle bowls, to the front door of a new five-star Intercontinental Hotel, next to a new three-block-long shopping mall.

When people ask me why China needs all the scrap metal Americans send to them, I wish I could show them the view from my hotel room that day. Twenty stories below is that shopping mall, as big as anything I grew up visiting in suburban Minneapolis. It required steel for the structure, copper and aluminum for the wiring, brass for bathroom fixtures, and stainless steel for all of the sinks and railings. And that's just the start.

Then there's this: on the other side of the mall, in all directions, are dozens of new high-rises—all under construction—that weren't visible from the subway and my walk. Those new towers reach twenty and thirty stories, and they're covered in windows that require aluminum frames, filled with bathrooms accessorized with brass and zinc fixtures, stocked with stainless steel appliances, and—for the tech-savvy households—outfitted with iPhones and iPads assembled with aluminum backs.

No surprise, China leads the world in the consumption of steel, copper, aluminum, lead, stainless steel, gold, silver, palladium, zinc, platinum, rare earth compounds, and pretty much anything else labeled "metal." But China is desperately short of metal resources of its own. For example, in 2012 China produced 5.6 million tons of copper, of which 2.75 million tons was made from scrap. Of that scrap copper, 70 percent was imported, with most coming from the United States. In other words, just under half of China's copper supply is imported as scrap metal. That's not a trivial matter: copper, more than any other metal, is essential to modern life. It is the means by which we transmit power and information.

So what would happen if that supply of copper were cut off? What if Europe and the United States decided to embargo all recycling to China, India, and other developing countries? What if, instead of importing scrap paper, plastic, and metal, China had to find it somewhere else?

Some Chinese industries would substitute other metals for the ones that it couldn't obtain via recycling—that's technically doable in many cases—but for some applications (like the copper used in sensitive electronics) substitutions are not possible. That leaves mining. To make up the loss of imported scrap metal, there'd need to be a lot of holes in the ground: even the best copper ore deposits require one hundred tons of ore to obtain one ton of the red metal. What would the environmental cost of all that digging be? Would it exceed the environmental cost of recycling the developed world's throwaways? What's worse?

*   *   *

In October 2012 I drive north on Minnesota's Highway 53 into the so-called Iron Range, which once supplied the American steel industry with some of the world's purest ore. As I approach Virginia, Minnesota, I begin to see the high, looming walls of dirt excavated from pits as deep as 450 feet, and as wide as 3.5 miles. They look like crater walls from the highway, left by meteor impacts and defining the landscape for miles. If you climb one (I did), you'll look out at a lifeless gray moonscape. This is what's left behind when steel is made from iron ore, and not scrap metal.

I continue north for nearly an hour and then take a right turn just outside the town of Ely, onto Highway 1. It's beautiful out here, green, lush, and uninterrupted. I see only two other cars on the road for the first ten miles; I stop my car on bridges over the shimmering blue Kawishiwa River without fear of being hit; I close my eyes down by the water, the only thing cutting the heavy blanket of silence the individual lapping waves.

I follow directions given to me earlier that morning and take a sharp left on to Spruce Road. There, at the intersection, is a bumper-sticker-festooned minivan that belongs to Ian Kimmer, staff member with Friends of the Boundary Waters, a group that aims to protect, preserve, and restore the federally designated million-acre Boundary Waters Canoe Area Wilderness (BWCAW), one of the largest unspoiled regions in the United States.

Ian has a big job. From the time the BWCAW was established in 1978 until now, the communities that surround it have expressed considerable hostility to the idea of an unexploitable wilderness in their midst. From their perspective, wilderness inhibits growth and the resource extraction industries that their towns and families were built upon. So far, they haven't made much progress in turning back or damaging the mostly pristine status of those million acres. But that's likely to change, and the single factor responsible for the shift is one that scrap-metal men know well: the price of copper.

For decades, geologists, mining companies, and miners have known that the land around the BWCAW contains deposits of copper ore. But those ore deposits are of such low quality that nobody could figure out how to mine them profitably. Then, in the 2000s, China entered the market for copper. What had once been worth 60 cents per pound became an occasionally $4-per-pound commodity, and a low-grade, unprofitable ore deposit became a mother lode that mining executives speculate might be the largest untapped extractable copper reserve in the world, worth around $100 billion.

Ian shakes my hand, takes a seat in the front seat of my Saturn, and sends me down the rutted dirt lane that is Spruce Road. On the left side, he notes, is the BWCAW. On the right, he says, pointing, is where the mining companies are doing test drilling.

"It's that cut-and-dried?" I ask.

"Yep." He asks me to stop, and we walk up a hill. Near the top, we reach a crumbling gray and red rock outcropping. It contains copper ore, he explains, as well as something called sulfides. When rain or snow comes into contact with sulfide ore like this, Ian explains, it produces caustic sulfuric acid. "That's why the rock is so crumbly."

Ian points at the base of the outcropping, where a several-foot-long streak of dirt is completely devoid of vegetation. "That's where the acid leaches out and down the hill," he explains, killing the vegetation. The phenomenon is not unique to northern Minnesota. Sulfide ores are mined around the world, and the leftover rock—the tailings—have become a long-standing environmental problem, contaminating rivers and lakes, and killing vegetation and the wildlife that depends on a clean environment.

According to Twin Metals, the mining company that controls the rights to the ore on this side of Spruce Road, Ian and I are standing atop 6.2 million tons of copper, 2 million tons of nickel (used to make stainless steel), and some of the world's richest untapped precious metal reserves outside of South Africa. Twin Metals hasn't received the permits to mine, yet, but if and when they do, each ton of copper will require the processing of as much as 100 tons of ore. Multiply 100 tons of sulfur-bearing ore by the 6.2 million tons of copper beneath my feet, and the scale of the problem becomes epic.

What will happen to the ninety-nine tons of sulfite rock once the copper has been extracted from it? Some will go back into the ground, Twin Metals claims, but an unknown percentage of those billions of tons will need to remain on the surface, exposed to rain and snow.

But that's not the only surface impact of this proposed project. Twin Metals is promising an underground mine—an "underground city"—using a method called "block caving." Superficially, at least, block caving sounds like a great compromise: the miners get the ore, and the wilderness remains untouched. But that's not how things work in reality. At some point, the surface will subside into all of the space left behind by the excavated ore, leaving a landscape substantially different from the one that was there before the mine. Rivers and creeks might be redirected; new lakes might be created. But that's the thing: nobody knows for sure. The one thing everyone knows, though, is that the unique character of this natural landscape will forever be altered.

Ian and I get back into the car, and he directs me down Spruce Road and an in-progress logging operation just off the BWCAW boundary. Trucks are loading freshly cut logs onto flatbeds, leaving behind little more than scrub. But Ian wants me to look past the logging, to two chest-high pipes painted red and sticking out of the ground like pins. "That's a test drilling site," he tells me. "There's hundreds of them all over the place. They're looking for the richest places to run the mine."

No Chinese company is involved in the Twin Metals project (the company is a joint venture between Canadian and Chilean firms), but Chinese demand is what makes the mine a virtual certainty. While Twin Metals investigates northern Minnesota, the Chinese are already digging some of the biggest and most controversial copper mines in the world today. In Afghanistan, the Aynak mine threatens ancient Buddhist sculptures. In Burma, a copper mine run by the Chinese military is destroying ancient farmland and causing mass protests.

Let me be clear: a doubling of U.S. copper scrap exports to China wouldn't halt this destructive trend. But it might just reduce some of the demand for that virgin copper.

In any event, when it comes out of the ground, all of that Chinese-mined virgin copper will have competition—from imported scrap metal, as well as from the scrap metal that the Chinese are generating in greater volumes at home. But cut off access to imported scrap copper, and the demand for mined copper will only grow—including the demand to allow mining in more places like Spruce Road.

Foshan, China, is the living, breathing alternative to the mine that will one day be dug somewhere near Spruce Road. It's not the cleanest industrial town I've ever seen, but unlike Spruce Road and its test drilling sites, it doesn't leave me with a feeling of intense personal loss. If anything, I always leave Foshan energized.

* * *

For the last two decades, much of the U.S.- and European-generated scrap metal exported to China flowed into Foshan, home of the Fontainebleau Hotel. But these days, if you're riding on the elevated highway that cuts through and above most of Foshan, you won't see any piles of metal, much less the smoke of burning wire and unvented furnaces. The people who live in Foshan's expensive new high-rises won't tolerate it. Instead, you'll just see under-construction buildings and long strip malls filled with restaurants and small workshops that sell construction-related supplies.

These days you need to turn off the highway, down the narrow city streets, and then into the even narrower lanes and alleys of Nanhai. The buildings are one and two stories high, and every one sits behind a high brick wall. But if you're lucky or—even better—invited, a gate will open here or there, and you'll see piles of baseball- and golf-ball-sized metal chunks; neat stacks of baled-up wire; machinery that takes fist-sized chunks of shredded automobiles and sorts them by size; and workers slowly combing through those same chunks, sorting them by metal type. It's a cleaner and wealthier Foshan, where worker salaries have quadrupled in a decade and many of the earliest and biggest recyclers sit on fortunes worth hundreds of millions.

For all of the cosmetic improvement, one thing in Foshan won't soon change: the hand labor of Chinese workers is essential to recycling the wasted luxuries of American and other developed world consumers. In 2011 I visited a yard where men dismantled old aluminum deck chairs imported from somewhere warm and vacation-like. Over to one side was a pile of the blue and white nylon stripping that once hung between the metal frames (later to be sold to a plastics recycler), and a woman who spent the evening cutting it away from the chairs. On the opposite side of the pile were men with chisels and pliers, busy breaking away the steel screws, fasteners, and hinges that "contaminated" the more expensive aluminum. Nearby, a similar process was under way, with aluminum screen doors hung with steel mesh that needed to be removed. The act might look mindless, relentless, and even dehumanizing, but from a business standpoint it's pure profit: aluminum contaminated with steel is all but worthless, a mixed metal that can't be sent to any furnace for remelting. But separated? Depending on the market, the aluminum might be worth $2 per pound.

Back at the Foshan Intercontinental, Joe Chen, a diminutive and gracious Taiwanese-American scrap man in his early seventies, picks me up in his chauffeured Mercedes. I've been invited to join him at a dinner he's hosting for several Mexican scrap exporters, and we glide through Foshan on the way to meet them. Living standards and wages in Mexico aren't much better than China's, but China has an advantage over Mexico: it's growing. So Mexico, poor as dirt, sends its scrap to the factories of China.

Joe understands the dynamics of this trade as well as anyone in the world. In 1971 he started traveling the United States, cold calling for scrap to send to scrapyards owned by relatives in Taiwan. "I flew, I drove. I went to yards without an appointment, and a lot of times I got thrown out. Today we are here, and tomorrow we are in the next state."

He specialized in low-grade scrap: insulated wire that needed to be stripped or burned, scrap radiators that had to be separated into aluminum and copper components, and loads of motors, water meters, and other metal-rich devices that had to be busted apart by hand to free up the constituent metals for sorting. It was the sort of scrap that used to be processed in the United States (and on my great-grandparents' basement stairs) until rising labor prices made the practice unaffordable, and environmental crackdowns shuttered the refineries and smelters that could do it chemically. By the time Joe started scrapping, much of that scrap had nowhere to go in the United States—except the landfill.

Joe's export business was so good that in the early 1980s he had the means to establish his own scrapyard in Kaohsiung, Taiwan, under the name Tung Tai. But Taiwan too was evolving, and as incomes rose, the public and its government became increasingly intolerant of the burning and dumping associated with the scrap industry. Meanwhile, as Taiwan's economy developed in the 1980s, $100-per-month workers became $500-per-month workers who were ready to join the middle class. "You couldn't find the workers anymore," Joe tells me. "They didn't want to do it!"

Joe realized that if he didn't find new markets, he'd own a business rich with suppliers of low-grade scrap across the United States, but—once again—nowhere but an American landfill to ship it. So he started thinking about China. It wasn't such a stretch: other Taiwanese industries that couldn't afford to operate in a more expensive Taiwan were starting to move there.

For two years, Joe searched fruitlessly for a Chinese local government partner or patron. Then in 1987, just as he was close to giving up, a delegation from Zhuhai, a port city in Guangdong, turned up in the United States and needed some help getting around. Joe was based in California, and he was more than happy to help. As it happened, one of the delegation's members was the "owner" of a large, government-owned scrapyard in Zhuhai. He'd heard Joe was in search of a place to import and process scrap, and after a week of being shown around the United States by Joe, he made Joe an offer. "You can have—you can rent my yard. Receive material there." Joe shrugs as he recounts the offer to me. "Zhuhai was my first yard."

It was 1987, and though China allowed private investment in the economy, outsiders were well advised to find somebody who could help ease the passage. "You need[ed] a relationship with the government at that time," Joe explains. "Without that you could not come." It wasn't just a matter of not being able to set up a yard, either. At the time, China didn't have any environmental regulations related to the import of scrap metal, nor did it have customs officials trained in the art of assessing a duty on scrap metal. In the absence of regulation, you needed somebody who could say, *I am the regulation, and here's your approval.* "Twenty years ago, nothing—no regulation, no customs tariff. I bring it in, they decide how to charge me. It's metal, copper—they don't know. They don't know how to charge me." The government was interested in jobs, presumably; the owner valued "rent"; and Joe wanted somewhere to process all that U.S. scrap he was collecting. If any one of the three links in this chain failed, then all that scrap was bound for a U.S. landfill.

At its peak Tung Tai's government-leased yard employed a breathtaking three thousand workers and imported five hundred containers per month of low-grade copper-bearing scrap like motors and insulated wire. The motors, Joe tells me, were purchased for two cents per pound, and contained copper worth thirty times that amount. Labor was just as cheap—less than a dollar per day. All the while, the market for scrap— and especially copper scrap—did nothing but grow. Between 1985 and 1990, China doubled its production of copper from scrap metal, to 215,000 metric tons per year, accounting for 38 percent of all copper produced in China, according to data compiled by the China Nonferrous Metals Industry Association. If Joe Chen was really bringing in five hundred containers per month, he might very well have been responsible for close to 10 percent of that supply in the late 1980s.

Joe was proud of Tung Tai's Zhuhai yard. As he saw it, the yard solved two important problems: it provided a place for Americans to recycle things that couldn't be recycled in the United States, and it employed thousands of Chinese. So in 1990 he invited international media to visit. "It's thousands of tons of scrap every year in the United States," he told Dan Noyes of the progressive *Mother Jones* magazine. "And the United States has got to find a place to dispose of it."

Noyes didn't disagree. His article described "discarded batteries, electrical motors, copper wire, even used IBM computers" scattered over Joe's yard. But unlike Joe, Noyes didn't see anything commendable about how Joe was handling the scrap. Rather, he saw wire fires, burning transformers, and a giant Tung Tai trash trench. Rather than expressing gratitude and admiration to Joe for taking all of these troublesome items off the hands of wasteful Americans, Noyes was indignant at the negative health, safety, and polluting effects of Chinese recycling methods. "From atop the factory's administration building," he wrote, "the scene was reminiscent of a prison chain gang."

Joe Chen, too, was bothered by the pollution (and he was quoted as saying so in *Mother Jones*), but he resolutely declined to blame himself. Instead, he pointed his finger at wasteful Americans and—perhaps unwisely—the people who allowed him to operate in Zhuhai in the first place: "Right now I've got the feeling the government [in China] only cares about the money. I don't think they realize the problem yet."

Predictably, the relevant authorities quickly recognized that their problem was Joe, and shut down Tung Tai's Zhuhai yard.

It was a rough period for Joe. "I think I talked too much," he tells me in the midst of a 2009 visit during which he decides it's time to talk about his moment of media notoriety (later, he offers a second assessment of the period: "Oh my god oh my god oh my god"). But in the long run it didn't matter: Joe now has several China-based yards and as many tons of U.S.-based scrap as he can handle. The "stuff," as Joe characterizes it, has to go somewhere, and he believes China is the best place.

When he invites me to visit his Guangdong scrapyards, Joe makes a point of showing me things easy to hold against him—like the worker dorms. "If I show you the best, then I must show you the worst. But if I show you the worst, then I must show you the best." So I walked through steamy dorms where the only personal space allotted to workers is the space inside their bunks. Those bunks, it must be noted, are in rooms that lack air-conditioning in the tropical Guangdong summer. Joe realizes this, but makes no apologies: "The conditions I give them are ten times better than what they'd have back home. In Hunan [Province] they'd be sleeping twelve to a room, sometimes to a bed. And they wouldn't be having eight-course meals." Later he makes me an offer: "You don't believe me? You can take my car, and I'll have my driver show you!"

I don't take him up on the offer, but I know what he means. The life of a rural Chinese villager is hardly bucolic. Homes are cramped, lacking in privacy, and often without plumbing. Depending on circumstances, meals are simple, and surely not as varied as those served in Tung Tai's kitchens (and yes, I've seen the eight-course meals). Rather than spending days sorting scrap for wages, villagers spend days in fields, picking crops for subsistence. Is one better than the other? I've never lived in either circumstance, so I'm not about to guess. But one thing I know is this: in the 2000s there was no shortage of laborers available to China's scrapyards. They lined up in the mornings, hoping for work, fresh from farming villages in the provinces. They could have stayed home; they could have gone to work in traditional factories; instead, they chose to work in scrapyards.

Why? The money. A chance for a future. Most of the money earned by those laborers was sent home, often to pay school fees for kids left behind.

Is the work safe? Sometimes it is, sometimes it's not. Breathing the smoke that rises off a pile of burning wire is not safe; neither, for that matter, is it safe to breathe the leaded fumes that come off a computer circuit board when it's exposed to flame. But most of what happens in a Chinese scrapyard is breaking and sorting. Burning, despite two decades' worth of exposés by environmentalists and journalists, is a very small and declining part of what happens in China (Africa, and to a far lesser extent, India, still burns).

In the early 2000s, I saw workers in little more than T-shirts, cotton slacks, and sandals working around open furnaces; I saw other workers using cutting machines and acetylene torches with their bare hands; and even today I'm not surprised to see scrapyard employees going about their work in flip-flops. Hard hats and safety glasses, respirators and work gloves, are as uncommon in most Chinese scrapyards as kosher hot dogs. Anecdotally, at least, injuries are common. Unfortunately, China's employers aren't under any requirement to report workplace accidents, so we really don't know how common they are.

Will China's scrap industry become more safe over time? Probably. But even in the United States, where workplace safety regulations are among the most advanced and best-enforced in the world, the scrap industry is still a leading source of workplace accidents. That's not for lack of trying: the industry's leading trade associations expend an inordinate amount of time, energy, and money on safety-related training. But one simple fact remains: cleaning up someone else's garbage is an inherently dangerous business. The best solution—really, the only solution—is to stop throwing away so much stuff. Every old piece of plumbing, every used computer, is just another opportunity for someone to be injured.

But for all of the risks, there are still opportunities, and in my travels I have yet to come across a country, a region, where recycling is in decline. As resources become more scarce, the demand for people to extract those resources becomes ever greater. It's an entrepreneurial opportunity for the small-time grubber, but in some ways it's an even bigger opportunity for the entrepreneur who figures out how to do business with that grubber. Nowhere on earth has the scale of that opportunity been appreciated, and seized, more readily than in southern China.

In 1980 China's central government designated Shenzhen, then a small fishing village located just across the border from Hong Kong, a Special Economic Zone. It was to be China's free-market reform laboratory, located far enough from Beijing to avoid contaminating the capitol with its bourgeois ideology, but close enough to Hong Kong to attract wealthy Chinese investors with experience in free markets.

Hong Kong had other advantages, first of which was its port. Then, as now, it was one of the world's busiest, attracting cargo vessels from around the world. For Joe Chen, and other early scrap importers, this was crucial. But Hong Kong was already a heavily trafficked stop for ocean shippers, and thus Joe knew he could move shipping containers of scrap metal right to the Chinese border on a regular, predictable basis. Once in Hong Kong, they'd be off-loaded, placed on barges, and run into China. Duties would be assessed—though back then the word *duty* was interchangeable with *kickback* and *bribe*—and the container would be trucked to a scrapyard. Connections were essential to the operation, including all of the corruption that Chinese connections entail, and thus most early scrap importers—like Joe Chen—really had no choice but to partner up with local governments with good port and customs relationships.

In the late 1980s, around the time that Joe was active in the doomed Tung Tai yard in Zhuhai, a government-owned factory in Shenzhen's Buji District approached him about partnering up. "They're making wire, copper wire, but they don't know how to get enough [scrap] for their factory," he explains to me in the back of his Mercedes. Then he shifts into the voice of his future partners. "'See if you can ship some copper to us, and we can do some business.'" Joe was more than happy to fill their scrap copper needs, and twenty years later he still maintains the partnership.

On the day that I visited the company, a handful of workers were busy stripping wire and breaking up window frames on a vast, concrete-surfaced lot in the shadow of the multistory wire factory. The cleaned-up wire was wheeled into the factory, while other metals were bundled up and readied for shipment to small factories around Shenzhen. This latter business is the sort of thing that small scrapyards have done since the Industrial Revolution. But in China, where access to raw materials like metals was tightly controlled by state-owned and -run monopolies, a small-scale retail scrap metal business was nothing short of a commercial revolution.

Think of it this way. If you were a Chinese engineer in 1980, and you had an idea for a new kind of brass ballpoint-pen ball, you had one choice: give the invention to your state-owned and -run employer. But say, just for argument's sake, you wanted to develop, manufacture, and market that invention on your own—could you do it? Shenzhen had opened for business, and technically you were allowed to start a company.

But even if you could obtain the financing to buy the simple machinery to make your new ballpoint-pen ball, where would you buy the brass? In 1980 raw materials—including metals—were a Chinese-government-owned and -run monopoly that primarily did business with other government businesses. They didn't sell bucket quantities to nobody engineers with a dream; rather, they sold huge quantities to massive, inefficient factories at fixed prices. More likely than not, they didn't even have a phone number you could call, or a sales counter you could approach. And why would they?

Enter Joe Chen, scrap metal importer. Joe was big in those days, but that's relative. Compared to a state-owned goliath, he was small-time and interested in one thing only: selling scrap to the highest bidder. So if you, a small-time engineer, wandered into his yard in search of a few buckets of brass with which you could make ballpoint pen balls, he was going to sell it to you—so long as you had the cash. And sell he did, all day long, to scrap buyers large and small, each in his own way building businesses, products, and buildings across Shenzhen and the Pearl River Delta. Unlike the customers of China's big state-owned metal companies, the buyers of imported scrap metal (from importers) were small, entrepreneurial, and new. They were also revolutionary: no longer would China's entrepreneurs be denied the materials necessary to make their products. Scrap was eroding the key choke point of China's centrally planned economy.

Consider the numbers, as provided to me by the China Nonferrous Metals Industry Association's Recycling Metal Branch (CMRA): in 1980, 22 percent of China's copper production was fed by scrap metal. By 1990 that number had risen to 38 percent, and by 2000 it had reached 74 percent. The growth is due to a number of factors, not least of which is China's ever-growing appetite for copper and its limited reserve of copper ores to feed it. But what's important—what's truly affected the Chinese and global economy—is that the 74 percent share in 2000 represented, largely, entrepreneurs feeding raw materials to other entrepreneurs (today, the number is closer to 52 percent, due to a number of factors, including the high price of scrap compared to virgin copper). Meanwhile, the top five Chinese scrap-importing provinces conform perfectly to China's top five provincial GDPs, starting with Guangdong. Entrepreneurship and scrap aren't the only reasons for that correlation, but they certainly played a role.

By the late 1990s, Shenzhen and the cities around it had become the world's leading manufacturer of nearly everything—auto parts to Barbie dolls. It had become the feared and celebrated Workshop of the World, the place where factories relocated from developed countries in search of cheaper labor and weaker regulation; the place where a farmer could become a titan of Chinese industry, a middle-class homeowner, or at least an employee in a factory where the work is better paid than the subsistence living earned in a rice paddy. Today it remains one of the most concentrated manufacturing zones in human history, the place where everything is made.

But that was only half of the story.

By the late 1990s Shenzhen and the cities around it had also become the world's leading importers of scrap metal, paper, and plastic. They had become, quietly, the Scrapyard to the World, a place where wealthy countries sent the stuff that they couldn't or wouldn't recycle themselves; a place where former farmers took that stuff, made it into new stuff, and resold it to the same countries that had exported it in the first place.

No studies exist on the role that scrapyards play in fostering entrepreneurism in developing countries. But plenty of former farmers turned manufacturing millionaires in southern China will happily name the man—and it's always a man—who sold them that first load of scrap metal. Over ten years of hearing those stories, and those names, I've yet to hear anyone mention the name of a sales manager at a state-owned company.

It's important to note that the story of China's development into a manufacturing—and recycling—behemoth isn't a new one. In a very real sense, this has been going on for a long, long time.

Take, for example, North America.

In the early nineteenth century, mechanized papermaking arrived in the United States. The market was wide: increasingly educated Americans were reading more newspapers, consuming more books, and writing more letters. To feed this demand, America's papermakers relied upon old rags—mostly linen—for high-quality, low-cost pulp. Unfortunately for the papermakers, however, Americans simply couldn't save enough rags to meet the demand for printed material.

So America's enterprising papermakers—and its entrepreneurial rag traders—made a very contemporary choice: they looked abroad to the more wasteful economies of Europe for their raw materials. According to data collected by the historian of American waste Susan Strasser, in 1850 Americans imported approximately 98 million pounds of scrap rags from Europe. Twenty-five years later the United States imported 123 million pounds of scrap rags, most of them from Victorian England.

It bears repeating: scrap rags are not clean rags. Rather, they're covered in filth of various types, including industrial, medical, and household. Nonetheless, as repulsive as one might find a barrel of used linen rags transported from England, nobody in nineteenth- or early-twentieth-century North America railed against the Victorians for "dumping" their "waste" on the still-developing economies of the former colonies. Rather, those Americans who bothered to pay attention to the trade viewed it as a necessary and economical (if occasionally distasteful) way to meet booming U.S. demand for printed material. As for the Victorians, the only objection might have come from England's own papermakers, who found themselves in a fierce and expensive battle for rags with the Americans.

Rags were not the only material that nineteenth-century Americans lacked in abundance. In the 1880s America's booming steelmakers started using open-hearth furnaces that could be fed with scrap iron and steel. Demand was huge: railroads and other infrastructure required vast amounts of steel. The U.S. was not yet scrapping its old infrastructure, however, so it looked abroad to Europe, once again, for raw materials. According to data culled by Carl Zimring, U.S. imports of scrap iron and steel grew from 38,580 tons in 1884 to 380,744 tons in 1887—a tenfold increase during, not coincidentally, a railroad building binge.

Those steel scrap imports would decline during the early part of the twentieth century, as Americans became largely self-sufficient in iron and steel scrap—throwing away as much as they consumed, in effect (all the while, iron ore remained the primary means of making iron and steel). Then, in the years leading up to World War I, Americans began to *export* modest amounts of steel scrap, mostly to Europe. This wasn't a great shift—it would be two decades before the scrap steel export trade became a really big business—but it was a sign of a maturing, less desperate industry. Indeed, at the same time that the United States was exporting steel, it was importing it too, suggesting that smart traders had figured out how to take advantage of global markets, rather than just local ones.

Expanding trade meant bigger problems, including disputes with local trading partners, foreign trading partners, and—most significantly—governments. So in 1914 the first U.S. trade association for the scrap industry, the National Association of Waste Material Dealers (NAWMD) was formed, and three years later—in the midst of World War I—the membership established an Export Committee (later renamed the Foreign Trade Committee). The official bulletins of the NAWMD (archived with the Institute of Scrap Recycling Industries, a direct descendant association, in Washington, DC) suggest that the nascent group had three major concerns: topping the previous years' annual banquet, promoting trade between members, and resolving disputes with customs and tax authorities.

The vast majority of the business that America's scrap industry did in the early twentieth century was local. Nonetheless, American scrap dealers, then and now, were keenly aware that Americans threw away much more than they could recycle. Thus, a short item in the September 20, 1919, bulletin of the NAWMD noted that at a June meeting of the Foreign Trade Committee, "It was further suggested that an effort [be made] to see that a list of the members of the National Association of Waste Material Dealers be distributed in some way throughout foreign countries with the idea of opening up markets for members."

It wasn't just American scrap associations that were thinking along these lines. On September 25 the Scrap Iron, Steel, Metals, and Machinery Merchants' Association of Manchester, England, sent a copy of its membership list to the NAWMD and asked if the Americans might reciprocate; and on October 19 the Wholesale Woolen Rag Merchants Federation for Great Britain and Ireland did the same. But it wasn't just Europe that was interested in American scrap—and that revelation seemed to catch even the grizzled leadership of the NAWMD by surprise. The September 4, 1919, the NAWMD bulletin included the all-caps headline JAPANESE HOUSE WANTS TO DO BUSINESS WITH THIS ASSOCIATION. Though it lacked an exclamation point at the end, it's not hard to imagine the leaders of the association mentally adding one after reading the following note:

Gentlemen:
We are indebted to the Chamber of Commerce of your city for your esteemed name.
    Most of our Directors are members of the Chamber of Commerce here, and we are dealing in the following lines:

Wool Stock, Wool Waste, Paper Stock, Cotton Waste, Cotton Rags, Waste Rubber, Burlap, Old Bags and Old Newspaper.

We are desirous of opening business relations with reliable large firms among the members of your association. We shall be much obliged if you will kindly introduce us to them.

Yours Very Truly,

THE JAPAN & CHINA RAW MATERIAL CO., Ltd.

T. Sasaki, Managing Director

Sakaemachi-Dori Rokuchome,

Kobe, Japan

The Japanese solicitation didn't mention scrap iron and steel, but within a decade Japanese steel mills had become the leading customers of American steel scrap exporters—by far. In 1932, for example, the United States exported a total of 277,000 tons of scrap steel, worldwide. Of that, Japan accounted for 164,000 tons. But that was nothing: over the course of the next eight years, resource-poor Japan's imports of American scrap steel accelerated to meet the demands of its steel-hungry military, reaching an astonishing 2.026 million tons in 1939, a year in which the United States exported 3.577 million tons of scrap steel, total. Though the practice was legal, it was at best morally tenuous: in 1939 Japan was two years into a brutal occupation of China—for which war crimes trials would be held—a fact not unknown in the United States, to be kind. Likewise, in 1938 U.S. exporters sent 230,903 tons of scrap iron and steel to Germany—long after its racial policies were well known.

The free-market-inclined U.S. scrap industry may not have been roused by these disreputable episodes, but the Chinese-American community was. In 1939 and 1940, they organized protests at docks where scrap metal was being loaded for Japan. The American scrap men weren't moved to restrict their trade (or show up for the protests), however, and the exports continued until President Roosevelt used his administrative authority to prohibit U.S. scrap exports to Japan and Germany in July 1940. The Japanese, undeterred, turned to Central and South America to meet their military-driven demand.

That stopped it for a while. But only for a while. After World War II, the exports started all over again, especially to Japan, and then—in

Bales of Christmas tree lights at L. Gordon Iron & Metal in Statesville, North Carolina. *All photos by the author except where otherwise stated.*

Bales of imported U.S. Christmas tree lights, ready to be recycled at Raymond Li's Christmas tree light recycling factory in Shijiao, China.

Li's recycling system uses flowing water and a tilted, vibrating table to separate heavier copper from light-weight insulation. Similar systems exist across China. This one runs in Ningbo.

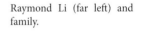

Raymond Li (far left) and family.

Cousin Yao (left), designer of Li's Christmas tree light recycling system, opens a several-thousand-pound sack of copper recycled in the factory.

Workers pull plastic from almost-sorted paper at Waste Management's Houston Material Recovery Facility. In the background, bales of sorted paper are stacked for shipment.

A bale of sorted paper in front of cages that hold sorted tin cans, juice boxes, and plastic bottles after they've been shot from the sorting line.

Bales of sorted, recyclable plastics at the Houston Material Recovery Facility. The bales in the center of the image are packed with curbside recycling bins that fell into trucks while being emptied.

Household recyclables in China are collected and sorted by peddlers, then sold to recycling centers. At this Beijing plastics recycling center, entrepreneurs rent space and hand-sort bottles into different plastic types.

The family-owned scrap warehouse where I learned the business.

My grandmother was the daughter of a Russian-Jewish immigrant junk peddler. Here she's seated next to the cash register and truck scale monitor at my father's scrapyard, doing what she liked best: buying scrap.

Inside the scrap warehouse. Rarely did a box of junk—as my grandmother called it—remain more than a week.

A worker at the Rama Paper Mills outside of Delhi, India, separates cardboard from white paper in textbooks and notebooks imported from the United Arab Emirates.

The result is higher-quality recycled cardboard and higher-quality white paper.

The sample wall at OmniSource's Fort Wayne wire chopping plant may be the world's most complete collection of the means by which Americans have transmitted power and information over the last half-century.

Pieces of 99.5-percent-pure copper, recovered from wire recycled at OmniSource's wire chopping factory. *Image—and hand—courtesy of Christine Tan.*

A 2002 image of the Tung Tai scrapyard in Foshan, China.

Joe Chen, Tung Tai founder and president, is one of the pioneers of the Chinese scrap industry.

One of the worker dorms at Tung Tai in 2002. "If I show you the best, then I must show you the worst," Chen told me.

A deceptively quiet lane in the Nanhai District of Foshan, China, likely the most concentrated recycling zone in the world. High walls hide hundreds of businesses devoted to the recycling of imported scrap.

A characteristic large-scale Foshan recycling business devoted to the hand-sorting of imported shredded automobile scrap.

The view from the control tower at the Yantian International Container Port in 2005. Approximately 10 percent of the containers hold imported scrap paper.

Bales of imported U.S. scrap paper at a mill near Yantian.

A container of brass "Honey" scrap shipped from Dubai is unloaded in Jamnagar, India.

Loose Honey.

After sorting, the metal is melted and cast into brass plumbing valves.

Workers sort and polish the finished valves. They'll eventually be packed and sold in Dubai.

A woman tends her used electric motor stall at the sprawling reuse market in downtown Taizhou, China.

Electric motors that can't be repaired and reused are dismantled into components.

Over the last thirty years, millions of electric motors have been exported from the United States to China for recycling. Chiho-Tiande in Taizhou is the world's largest motor processor. In 2010 it was listed on the Hong Kong Stock Exchange.

A worker at Taizhou Xinglitong Metal Industry Company dismantles motors attached to a machinery array imported from Japan.

Piles of sorted electric motor parts dismantled in Taizhou, China. Most of these parts will be purchased and used in rebuilt motors. Those which are too worn will be recycled.

At Net Peripheral in Penang, Malaysia, stacks of imported, used computer monitors from the U.S. await testing and refurbishment by technicians.

A worker buffs out the scratches in a used computer monitor. Later, it'll be placed in a case and shipped for sale, most likely in developing Africa.

Johnson Zeng searching for scrap at the OmniSource warehouse in Spartanburg, South Carolina.

Homer Lai, scrap metal importer based in Qingyuan, China, started out as a barber. He now imports the majority of the scrap that Johnson exports.

equally stunning volumes—to Taiwan. Somebody had the junk, and somebody wanted it.

Nothing changes.

During my decade in China, life went on in Minnesota. My father went in and out of treatment; my childhood home was sold; and my grandmother passed away. Still, looking back on all these years, the moment that pushed me most off-balance, the one that reminded me that I couldn't go home again, was the day I learned that the city of Minneapolis had bought the land beneath the family scrapyard.

The yard was where I'd grown up in some ways, where my grandmother and I had spent our sweetest moments, eating kosher hot dogs in the mornings, weighing trucks on the truck scale. She'd live a few more years after it closed, but she was never quite the same. Neither was I. True, my father reopened a small metals warehouse on the north side of Minneapolis. But it was too far for her to drive, and it didn't have any of the history, it didn't have any of the scrappy memories. In my mind it wasn't a serious business anyway, more a hobby akin to a backyard garden. I've set foot there six times, twice so that my grandmother could enjoy the sight of scrap metal before she died. As for my father, he still has the talent to work a few hours per day and earn enough to keep a small warehouse running.

Once in a while, in China or the United States, people ask me why—with all of my knowledge and contacts—I don't go into the scrap industry myself. "You could make a lot of money," they tell me. "You know so many people. It's such a waste." That's all true, I suppose. But what those who covet my contacts don't understand is that the *only* reason I know so many people, the *only* reason I have so many scrap-metal friends, is because I'm *not* in the business. The moment I started selling scrap, brokering scrap, buying scrap, all those people who talk to me—all those scrap-metal friends—would become scrap-metal competitors. Maybe I'll change my mind one day, but for now I'm just so much more interested in keeping them as friends. Losing them, it seems to me, would be a waste.

CHAPTER 5

The Backhaul

I was in junior high school when I realized that my father was shipping scrap metal to Asia. Best as I can recall, I was snot-nosed and righteous. How, I asked, can you make money doing that?

I don't recall the response that I received, but I suspect it went a little something like this: "The Chinese are paying more, shut up." In family scrap business terms, that's usually the end of the discussion. But it wasn't an answer—the answer—that I was seeking. That would only come many years later, as I gradually came to appreciate how tightly connected Chinese demand for American recycling is to American demand for Chinese goods. Fortunately, the global scrap industry recognized the connection much earlier than I did, and in doing so—and taking advantage of it—created a multibillion-dollar sustainable business model that stands as one of globalization's great, green successes. Ironically, it's a mostly untold success, despite the simplicity of the tale and the benefits that it's brought to businesses, consumers, and the environment alike over the last twenty years.

It's a steamy southern China afternoon in 2005, and I'm stepping into an elevator bound for the top of the control tower overlooking the Yantian International Container Terminals, or YITC, the second largest port in China, and the fourth largest in the world. This is an impressive achievement, especially considering that the YITC is less than a decade old when I visit. However, it's not a surprising achievement: Yantian is a district of Shenzhen, and the YITC is the place where the Workshop of the World exports most of its goods.

The elevator doors open, and I stride into a room that looks like mission control at NASA. At the front are giant screens beaming digital maps streaked by moving yellow lines representing ships as they sail in and out of the port. Below them are rows of technicians stationed at computers, charting itineraries and cargoes. But my eyes are drawn to the panoramic windows, and a checkerboard landscape built from tens of thousands—maybe hundreds of thousands—of twenty- and forty-foot-long metal shipping containers painted red, yellow, blue, and gray. The containers cover hundreds of acres, starting from the long piers and running right to the base of the tropical mountainside in stacks that rise as high as sixty feet. Here and there, a specially retrofitted forklift carries one down the canyons carved between the stacks; along the water, cranes lift the boxes from their stacks and place them delicately upon Empire State Building–sized container ships bound for the world.

The staff tells me that $147 billion in goods moved through here in the last twelve months, dating back to September 2004, carried in the equivalent of more than 13 million of these containers. I can't visualize those numbers; they're just too big. But the next number they provide is easy to visualize: only 10 percent of the containers held imported goods. The remaining 90 percent carried exportable goods.

The 90/10 split is the inevitable by-product of decades-old trade imbalances between China, a country that manufactures more and more, and places like the United States and Europe, which manufacture less and less. In 2005, the year I ascended the YITC control tower, China exported $243 billion worth of goods to the United States; in return, the United States exported $41 billion worth of stuff to China. That trade imbalance—or deficit, if you're an American—hasn't gone away, and today similar disparities exist for many of China's other trading partners, including the European Union and Japan. Like the United States, those countries don't export nearly as many containers of stuff to China as they import.

The sting of trade deficits afflicts many, but none more keenly than shipping companies. After all, they're in the business of moving containers from where people manufacture goods to where people buy

them. If trade is balanced—that is, if the United States and China manufacture the same volume of goods for each other—then containers going
to the United States with goods will return to China with goods. But
if the United States and the European Union aren't manufacturing anything that people in China want to buy, then shipping companies need
to figure out the cheapest way to move their containers back to China
ASAP so that they can be restocked with more goods to ship back to the
States. Shipping empty containers is one option, but not a very profitable one.

So what's a shipping company to do?

Step one, from a shipping company's point of view, is to hold a sale
and discount shipping rates to attract companies that might otherwise
avoid export due to the perceived or real high cost of shipping. Which is
precisely what has happened: the shipping companies have been dis-
counting so-called backhauls for decades. In early summer 2013, for example, the price of shipping a 40,000-pound container from Los
Angeles to Yantian was a paltry $600. Going from Yantian to Los Ange-
les, however, could cost four times as much.

Step two, if you're a shipping company, is to find an industry that (a)
generates millions of pounds of product per year, and (b) has customers
in China. Agricultural products like soybeans and wheat meet both requirements (as China grows richer, it eats more). But they tend to be
grown and collected in such large volumes that they're sent loose, in
the spacious hulls of giant so-called bulk ships. With that as an option, it doesn't make much economic sense to incur the expense of
packing soybeans into bags or boxes and then sending them in shipping
containers.

Which leaves, really, only one high-volume product to fill all of those
empty containers waiting to go back to China: scrap. And if you're looking for scrap that can fill up containers, then there's nothing better than
blocks of scrap paper the size of hay bales. Consider: in 2012, American
recycling companies harvested 45.45 million tons of old newspapers,
magazines, office paper, and boxes—four times the amount they generated of scrap plastic, copper, aluminum, lead, zinc, and electronics,
combined. But the best news, if you're a shipper, is this: the millions of
pounds of cardboard in which Chinese manufacturers pack goods for
export to the United States are in high demand by the Chinese paper

mills that make those boxes. In other words, that box your new Chinese-made television was packed in might have once been the box that your last laptop was packed in.

Round and round it goes.

In 2013 Americans exported almost 20.9 million tons—or roughly 45.9 percent—of the used paper and cardboard they harvested. Of that, the majority went to China in shipping containers that otherwise would have crossed the Pacific Ocean carrying nothing more than air. It was joined by millions of tons of recycled metal and plastic, all of which went—like that paper—on what amounted to the unused portion of a round-trip ticket from China to the United States, paid for by American consumers eager for Chinese-made goods. One way or another, the boat is going back to China, and the fuel to send it there is going to be burned, whether or not the ticket is paid for. So anybody—or anything—hopping on that boat is getting what amounts to a carbon-neutral boat ride to China (the weight of the containers is irrelevant: boats need ballast to balance in the water, and containers full of paper fulfill the function very well). Of course, the same cannot be said of the weekend recycler who drives the recycling down to the local county dropoff, burning gas all the way.

The difference in price between the two legs of the round trip varies by the ship, the time of year, and the port. But the critical fact for American scrap companies competing against Chinese companies is that the price of shipping to China is typically cheaper than shipping between geographically distant U.S. cities. For example, in late 2012 a container shipped via railway from Los Angeles to Chicago could cost as much as $2,400—or four times more than it would cost to send the same container to Shenzhen. In other words: U.S. demand for Chinese goods means that a paper mill in southern China can outcompete a Chicago-area paper mill for a shipping container of old newspapers in Los Angeles. That's the power of the backhaul—and American demand for Chinese-manufactured goods.

Perhaps one day the trade imbalances between China and the United States will disappear, and recyclers will no longer have a cost incentive to ship to China. But until that happens, the suspicious environmentally-minded recycler (such as the junior high version of myself) can take solace in knowing that less pollution will be generated sending a recycling bin full of paper from Los Angeles to China than sending it up to Seattle in an electric car. For businesses that can compute those fuel savings in financial terms, the profits can be substantial.

It's 5:45 A.M. in Los Angeles, July 2011, and Alan Alpert, president of one of those businesses, is speeding down an LA freeway. He drives a black BMW that looks freshly washed and—purposely or not—matches his black polo and black nylon running shorts. It's awfully early to go in to the office, he concedes, but he has to work out before his regular 7:30 A.M. meeting.

In his left hand is the steering wheel, and in his right is an electric razor that he's running along the left side of his face. Somehow, his eyes are focused on the road and his head is focused on the conversation he's having about used beer and soda cans (used beverage containers or UBCs, in industry parlance) over the car's deeply resonant stereo system. The other voice belongs to someone in the Central Time Zone, and that voice has some ideas about how more money can be squeezed from old cans. The deal is complicated, but by the time we reach the two-story Alpert headquarters, it's done, and the cans will soon be on their way to someone on the other side of the Rocky Mountains.

As Alan powers down the BMW, he explains something interesting to me. The price of shipping a pound of cans across the Rocky Mountains to a company that can remelt them into new cans is around 7.5 cents per pound—or roughly $3,000 for 40,000 pounds. But on the West Coast, at least, shipping to China via a giant cargo ship is much cheaper: around 1.5 cents per pound, according to Alan. In other words, that same 40,000 pounds can be shipped to Shanghai for $600—or $2,400 less than it costs to truck it over the Rockies. If, like Alan Alpert, you spend your days thinking in terms of hundreds of containers of old beer cans, a $2,400-per-container difference adds up to some real money.

Alas, there's a problem: China prohibits the import of scrap beverage cans due to concerns about the health and safety implications of importing the residue of old six-packs. It's a bizarre and seemingly random prohibition, especially for a country that allows the importation of far more dangerous scrap items, but there's little that Alan or anyone else can do about it. China has made its rules, and it takes time to change them.

"Any chance, you think, they'll lift that ban?" Alan asks as we leave the car.

"No, but . . ." In recent years, there's been talk among some of my Chinese government contacts that it's time to open China to more types of recycling, including beer cans that still have a little liquid in them. In part, they want to legalize what's already happening: I know of several Chinese importers who make a bundle buying cans from Thailand's resort islands and smuggling them into southern China.

Alan, however, isn't about to take that kind of risk. His company is too big and too well regarded in the industry to be caught up in smuggling old Bud cans.

There are a couple of cars in the parking lot, one of which belongs to J.J., Alan's buff personal trainer in a Bruce Lee T-shirt, who as he walks into the building with us mentions that he spent part of the weekend working out members of Muse. It's just after 6:00 A.M., but lights are already on in several of the windowless offices—especially those occupied by employees involved in arranging the shipping for the hundreds of containers per month that Alpert sends around the world. Alan proceeds directly to the small private gym.

One hour later, his workout complete, Alan climbs a flight of stairs and takes a right into the main conference room, where a dozen men, mostly in their thirties, are gathering around a conference table. Alan is dressed like they are: khaki slacks and a well-pressed long-sleeved button-down shirt (blue, in his case). A lamp with two fluorescent bulbs hangs over the table, providing just enough light to illuminate the meeting, but not enough to lend the impression that it occurs during daylight hours.

Alan takes a seat at the head of the table, leans back in his chair, extends his long legs, and looks over his shoulder at a large monitor that displays key financial data from around the world—including metal commodity prices reported by the major exchanges. One of the young men presses some buttons on a phone in the middle of the table, and a voice from Alpert & Alpert's Paris representative office comes on the line to say good morning, followed a moment later by the person in charge of Alpert & Alpert's risk management office in New Jersey. As Alan reminded me over dinner the night before, when a container of copper scrap can be worth more than $170,000 and can fluctuate by tens of thousands of dollars in the space of a day, there's a lot of risk to manage. Alpert & Alpert uses hedges—complex financial instruments pioneered on Wall Street—to take care of the problem. In other words: this isn't my father's scrapyard.

Alan turns to the table and looks at one of the more mature faces. "Take it away, Terry."

Terry Baumsten is in charge of Alpert's nickel, stainless steel, and high-temp (read: highly expensive) alloy business, and he gives a brief report on what the markets did overnight in Asia. "The three-month has moved up quite a bit since opening," he says, referring to the price of nickel on the London Metal Exchange, looking forward three months. "The stainless market is starting to percolate a little, with [Big Company A] looking for two hundred ton per month." I look up: that's an offer worth potentially more than $1 million per year. But Terry doesn't flinch, just goes on offering a range of prices that Alpert will be paying today for the metals under his authority. As he does, young traders reach for calculators, calculating the margins, the shipping, the profits. Later, they'll hit the phones to see if they can find someone who can fulfill the order among their wide range of contacts at scrap companies across the United States, and the world.

Alan takes a sip of water. "So what are you guys hearing out there?"

There's a brief silence; coffee cups are sipped from, early-morning shoulders are stretched. Then, at one end of the table, a throat clears and Jim Skipsey, Alpert & Alpert's director of purchasing, raises his eyes to the group. "I spoke to [Big Company B] this morning," he says, referring to a large South American scrap company. "The [name withheld] family is out. But they called this morning, looking for a container."

Skipsey isn't just a specialist in purchasing scrap, he's a specialist in purchasing scrap from Central and South American scrapyards. Thus ensues a brief discussion of the dangers faced by many prosperous South American scrap families, including the ever-present risk of kidnapping. Still, despite South America's risks, Skipsey—a Spanish speaker raised in Mexico—is an optimist about the region: "There's a lot of metal moving through Guadalajara."

Alan turns to the far end of the table. "Harvey?"

Harvey Rosen runs Alpert's considerable aluminum practice, and his deep-set eyes give him an intense presence. "You'll see an interesting number from Tsuji," he says, referring to Alpert's representative in To-kyo, and a price sheet that he's passing to everyone at the table. "Litho at one-oh-five." Meaning: Tsuji has somebody interested in buying litho-graphic printer plates of the sort used in publishing for $1.05 per pound. "I think we'll take a position," Harvey says, meaning, Alpert is going to sell the material and then try to buy litho based on that "interesting" price. "We can buy litho in the Midwest. We haven't done litho in Japan in a while."

Alan nods. "Did you get some feedback on mixed pucks from Tsuji?"

Mixed pucks are hockey-puck-shaped masses of compressed metal shavings created when, for example, a factory drills a hole into a piece of aluminum. Before Harvey can answer, Jim Skipsey jumps in to suggest some destinations. "[Big Company C in India] uses a specific alloy for pistons. So India could be a destination. Also—in Mexico, they could take it." Later in the morning he'll call around to see if he can close a deal.

The price, obviously, will be key to that closing. But the cost of ship-ping will be just as important. Mexico and India are both playing in the same global market, and are thus likely to pay within a few cents of each other. But shipping is another matter, and the difference in cost between sending a container to Mexico rather than India can sometimes be mea-sured in the thousands of dollars. Needless to say, all prices being equal, those mixed pucks will go where they can be shipped most cheaply.

Alpert & Alpert is a big and complicated global business, but it cer-tainly didn't start out that way. Rather, it evolved from modest circum-stances that wouldn't have felt unfamiliar to Leonard Fritz or the other American scrap grubbers of earlier generations. The two men who built it are located across the hallway from the morning meeting room in a large office overlooking railroad tracks, a busy street, and Alpert & Alpert's neatly organized seven-and-a-half-acre scrapyard. Raymond Alpert, a jocular and warm eighty-three-year-old who still comes to work four days a week (this is his office), is seated behind a large wooden desk. Next to him is Jake Farber, a svelte and handsome eighty-five-year-old with a wry smile and a quiet charisma.

Raymond and Jake joined the business in 1950, when Alpert & Alpert was still a modest-sized scrapyard of no particular note on the outskirts of town. On an average day, Raymond tells me, they had perhaps three hundred customers, most of them peddlers, many peddling amounts that amounted to no more than a few hundred pounds of steel. "We even had horses and buggies coming here in the 1950s," Raymond tells me. When I express surprise, he informs me that when he started, Los Angeles's paved roads dead-ended into dirt out this way.

For most of its midcentury history, Alpert & Alpert was a steel scrap company. It bought from peddlers and sold to the three steel mills then operating in the Los Angeles area. The other metals—copper and brass, in particular—were shipped to the Midwest, at considerable cost, due to a lack of factories that wanted them on the West Coast. That state of affairs, however, was due to change—though the change had almost nothing to do with anything happening in California or the Midwest.

In 1950 the U.S. government lifted a nearly decade-old World War II–era ban on scrap-metal shipments to Japan. It was a prime commercial opportunity for any U.S. scrap man with excess metal to export: Japan's postwar rebuilding effort was accelerating, and the island nation was desperate for raw materials with which to build highways, buildings, subways, cars, and goods to ship to the United States. In 1950 the States exported a mere 1,433 tons of steel to the island (total U.S. iron and steel scrap exports were 194,114 tons that year, with Canada accounting for 37 percent of that). But the trade grew quickly: in 1956, the United States exported 2.4 million tons of scrap steel to Japan, and in 1961 it sent an astounding 6.1 million tons. Today that would be an unusually large volume of scrap steel for U.S. companies to ship to one country; in 2013 the United States exported 17.5 million metric tons of scrap iron and steel in total, worldwide.

In the 1960s Raymond and Jake started sending some of their metal to Asia. This wasn't so unusual on the West Coast; even then, it was expensive to send scrap over the Rockies, especially as compared to sending it to Japan via one of the many freighters that made the circuit between the two countries. "Korea, Japan, Taiwan," Jake tells me. "Then we started going to Hong Kong when—" He pauses just briefly. "Those days, it was against U.S. law to sell any scrap to China. So we would send it to Hong Kong, and what they did with it I don't want to know. They probably shipped it to China." Or just as likely, they brokered it around Asia to markets that no American could hope to access. Vietnam, for example, was a market that had an interest in U.S. scrap metal, but trade embargoes prevented anyone but well-connected Hong Kong and Singapore brokers from making the deals.

By the mid-1980s, roughly half of Alpert & Alpert's metal was exported, likely making it the largest privately held exporter of nonferrous scrap metals on the West Coast of the United States, a trade worth millions of pounds and hundreds of millions of dollars. That exalted status was the result of many factors, but arguably none was more important than the insatiable American appetite for low-cost Asian goods. If Japan, Taiwan, Korea, and China could not export products to the United States, the low-cost backhaul shipping rates that still allow Alpert & Alpert to ship six figures' worth of metal to China for $400 simply wouldn't exist. More likely than not, they'd still ship to Asia, but the opportunity—and the profit—simply wouldn't be as grand.

Early on a Tuesday morning I take a seat across from Howard Farber, president of the Alpert Group and son of Jake Farber, in his first-floor office across the street from the 7.5-acre scrapyard that the company has operated for more than half a century. Like his father, Jake, Howard spent a significant amount of his early scrap industry career on the road in Asia, looking for new markets to which Alpert & Alpert could sell scrap. "I spent five weeks in Taiwan once," he recalls. "I started in Taipei and worked my way to Kaohsiung, and I stopped at every single ma-and-pop place."

Unprompted, he recalls that one of those "ma-and-pop" places was a foundry where hot metal was poured into molds. After the hot metal cooled, an elderly woman wielded a hammer to break off the excess metal that remained from the pour. "She's squatting down on the ground, wearing flip-flops and breaking this thing up. And the guy I'm with, the owner, he's like, 'I'd like you to meet my mother.'" Howard shakes his head. "There's no *way* my mom would be doing that."

As Alpert & Alpert's Taiwanese customers shifted west to China in the 1980s and '90s, so did Howard's travels. What he found in China was really no different from what his father had found in Korea, Taiwan, and—to a lesser extent—Japan in the 1950s: poor countries determined to industrialize quickly. Imported scrap—cheaper and easier to obtain than ores mined from the ground—was a key sustainable means of doing that.

Within a few years China was the biggest market by far for Alpert's metals. But Alpert & Alpert wasn't the only company to benefit: by 2000 China was the world's biggest importer of scrap metal and paper. Low-cost labor and lax regulation played an important role in that shift, but they weren't decisive factors by any means.

After all, then as now there were places where labor is cheaper than in China, and environmental standards even lower. Indeed, if labor prices and environmental standards were the sole determinants for where scrap (or waste) goes, Sudan—with labor rates well under $1 per day—would be the world's top scrap-metal importer.

So why isn't it?

The most important reason is that Sudan doesn't have many factories where scrap aluminum can be transformed into new aluminum and then remelted into new car radiators. Without such end markets—or the possibility of such end markets—there's absolutely no reason for a Sudanese to import $60,000 containers of scrap metal. In fact, the lack of such buyers means that the relatively small quantity of scrap produced in Sudan is actually *exported*, with much of it going to India and China.

But what about India? Why isn't it, and not China, the leading importer of scrap metal from the United States? On the high end, India's scrap laborers make perhaps $80 per month compared to the $250 per month that the cheapest of Chinese scrap laborers make. Equally important, India is home to a growing array of manufacturers in need of metal, including automobile makers. But despite all of those apparent advantages, India's scrap-metal imports from the United States are a fraction of China's. Why?

Howard leans back in his chair and tells me that, among the early office jobs he held before traveling abroad for Alpert & Alpert, one of the more educational was in the company's Traffic Department. It's an uninspiring name for what is the second biggest scrap-related expense—after scrap—at Alpert & Alpert: shipping. And it was in the Traffic Department, Howard tells me, that he learned a very important lesson about how and why scrap of any kind—metal, paper, plastic—flows to its final destination: "Scrap is gonna go to the place where the labor is cheap. That is correct. But if the labor's really cheap in India, and it's seven cents per pound to ship it to India, and it's two cents per pound to ship it to China—you know, unless the price is a whole lot better in India, it's going to China."

"And it's almost never the case [that the Indian price is better]," I respond.

"It's *rarely* the case."

Rather, it's usually the case that the Chinese price—driven by desperate demand—is better than India's. But even if it isn't, the price to ship from Los Angeles to Yantian is significantly cheaper than the price to ship to India's main scrap-receiving ports from the U.S. West Coast. The reason for this latter disparity is simple: India simply doesn't export many products to the West Coast of the United States. Until it does, the shipping companies aren't going to have much incentive to offer discount shipping rates for containers moving from Los Angeles to Mumbai.

However, India *does* export huge volumes of food and other goods to wealthy countries across the Middle East. Even more than their wealthy American, European, and Japanese counterparts, these Middle Eastern countries lack much of anything with which to fill those empty containers for the backhauls (oil, after all, moves in tankers, not containers). But, being rich, they do generate a lot of waste—in fact, per capita, they're much more wasteful than Americans, even.

So, no surprise, the top export, by volume, from Dubai to India is scrap paper and metal. Depending on the time of year, the cost of a backhaul to India can be as little as $200, and the time of passage can be as little as three days. A U.S. backhaul to India, by contrast, could cost seven times as much and require six weeks to traverse the Pacific and Indian Oceans. Other Middle Eastern countries—especially Saudi Arabia—generate even greater volumes of scrap that, taken together, are more than sufficient to feed India's still modest (compared to China's) but growing raw material needs.

This isn't the clean, crisp picture of waste dumping that exists in the West. Rather, it's something more complicated: the emergence of a truly global market in old goods. Africa exports vast amounts of scrap to China; China exports scrap televisions to South America; South America sends scrap wire to China. All of this globalized scrap, every last hunk, moves according to who wants it most, and who can ship it most cheaply. The proof of the phenomenon can be found everywhere, but for me it's never more clear than in Brass City.

*   *   *

It's late afternoon, August 2010, and dark thunderheads churn over Jamnagar, India. I'm seated in a late-model Jeep driven by Sunil Panch-matiya, former local cricket star turned successful brass scrap importer, and it's rocking like a boat as he negotiates the muddy potholes and cattle that disrupt the road. "There are fourteen factories in Jamnagar making tire valves!" he yells over the clatter. "The most in the world!"

I hold on to the door handle and nod. After three days in this outpost of 800,000 people in northwest India, it's been impressed upon me—repeatedly—that Jamnagar holds the world together. Brass belt buckles, brass pen clips, brass shoelace ties—Jamnagar manufactures them all, and in greater quantities than any other place on earth, and—this is what's important—they do it with scrap metal. That's why I'm here, in fact: for years I've been hearing about this mysterious place, where—depending on who you ask—between three and four thousand small workshops toil away, making the widgets of daily life from imported scrap brass.

Welcome to Brass City.

Sunil turns right, and we stop in a narrow courtyard surrounded by single-story concrete buildings streaked with soot and rain. A balding and slightly overweight man on the verge of middle age rushes from one of the doorways, crosses to another, and gestures for us to join him. So we rush out of the car, into the rain. As we dash for the doorway, I notice a small pile of mixed brass scrap in a corner of the courtyard: faucets, candleholders, vases—the kind of thing that my father used to buy on the scale in Minneapolis. Of all the ISRI specifications, the name chosen for this material strikes me as particularly apt: Honey. When you look at all that warm yellow brass in the rain, it looks like dirty gold.

I slide to a stop in a concrete room lit by a dark, greenish fluores-cence. Five thin, lithe men glance at me from their stations at sewing-machine-sized devices—*click-click-click*ing away—positioned around the room's perimeter. Then they go back to work. On the floor, I note, are piles of bullet-sized pieces of freshly machined brass. "They make two million air valves every year," Sunil tells me.

The paunchy middle-aged guy lights a cigarette and hands me his card, but asks that I not use his name. The company name is fine, though: Jayesh Impex Pvt Ltd. "We're done melting scrap for the day," he says. "So you can only see the manufacturing part."

"How much scrap do you use?" I ask.

"Fifteen tons of Honey, minimum, every month."

"Where do you get it?"

"Sometimes Sunil, sometimes somebody else."

I turn to Sunil: he gets it, I'm told, mostly from the Middle East. Depending on the condition of the local economy, and the global economy, Jamnagar's brass importers bring in between three and four hundred containers of Honey per month, most of which originates in the Middle East and Europe. Roughly estimated, that's 12 to 16 million pounds of faucets, vases, bits and pieces of manufacturing waste, and anything else made from yellow brass that made it into a scrapyard for shipment to this town up near the Pakistani border.

Sunil is a modest-sized scrap importer, bringing in less than twenty-five containers per month, mostly from Dubai. By contrast, the biggest Brass City importer is responsible for bringing in perhaps 30 percent of the scrap flowing into Jamnagar, and his sources are found throughout the Middle East, Europe, and even—occasionally—the United States. In any case, both Sunil and that big guy do the same thing: resell percentages of containers to smaller players like Jayesh Impex who lack the cash to buy full containers on their own. It's not unlike the role fulfilled by Joe Chen in Guangdong in the late 1980s and early '90s—except that Jamnagar's brass trade dates back to the early 1960s, at least. But in India, it seems, things move a bit more slowly. Still, Sunil assures me, Jamnagar is changing.

I walk slowly around the room, watching as workers run machines that transform small hollow tubes into air valves sold to bicycle tire manufacturers around Asia (China, mostly). Those tires, then, will be sold in bike shops around the world.

"Every valve is checked for quality by hand," the balding manager explains as he takes me into an adjoining room where two men spend their days pinching the ends of bicycle tire valves and dunking them into water. If bubbles emerge, the valve is defective and sent for remelting. If bubbles don't emerge, the valve is good for export.

Sunil tells me that the workers are paid between $60 and $80 per month—a price so cheap it hasn't been seen in China in a decade and, barring an unforeseen total economic collapse, won't be again.

"This is a hundred percent green business," explains the balding manager, referring to the valves on the floor, made from scrap recycled in an unfiltered, coal-fired furnace. "And it is growing all over Jamnagar. There are fifteen to seventeen valve makers just like us."

In the morning, Sunil drives me through teeming, winding, impossible-to-untangle downtown Jamnagar. I have very little experience in India, and so—to my eyes—it's a stream of colorful saris and turbans, and a wild mix of colonial architecture, temples, 1970s-era concrete box buildings, and motorcycles. Then, abruptly, we are driving down the straight muddy lanes of the GIDC (Gujarat Industrial Development Corporation), an industrial park built in the 1960s. Motorcycles, scooters, and the occasional bicycle line the road and the front of muscular fifty-year-old concrete buildings, one after the ugly other. As Sunil parks beside a factory wall, I notice that smoke—thin, but perceptible—hangs over the area.

"I got my start here," Sunil says as we step out of the car. He's a former cricket star, a fact that weighs upon him, and so he adds: "After cricket." In his mid-forties, he's a suave and movie-star-handsome presence in jeans and a silk shirt. He's not a natural scrap man, best as I can tell, but his family is in the brass business, and thus, in the end, so is Sunil.

I notice, through an open doorway, a pile of scrap and—beside it—two men using a handsaw to cut through a refrigerator-sized brick of zinc. One pushes, one pulls, and the dust falls onto a piece of burlap, later to be collected. It's a mind-boggling sight—but it's nothing that I haven't seen before. Four years earlier, in fact, I saw guys in Mumbai sawing television-sized bricks of copper by hand. "Brass is made from copper and zinc," Sunil reminds me. "But these bricks are too big to fit into Jamnagar furnaces. That's why they cut them."

"Can they live on that work?"

"They're probably migrants from Punjab," he tells me, then says a couple of words to them in a language I don't understand, and they nod. "They send money home. It's better than farming. I'm sure they're illiterate."

Sunil tells me that there are 1,100 brass companies in the GIDC, and as I walk the muddy streets, dodging rickety trucks carrying loads of old faucets and newly manufactured brass rods, I find no reason to disbelieve him or anyone else who's promoted Jamnagar to me. Every spare lot, every corner, is occupied by somebody with a burlap bag overflowing with yellow brass.

Down the muddy road, a potbellied man in white slacks and a blue-and-white-striped silk shirt emerges from a building, and when he notices me, the rare white face in these parts, he beckons with a wave. I'm in the habit of accepting such invitations, but I'm not sure about Sunil. I glance back at him, and the smile on his face is familiar, as if he and the man go way back.

"When I was starting in the scrap business, he and I were neighbors. C'mon."

The potbellied man's name is Pravinbhai Timbadia, and his company, Jai Varudi Enterprises, fills a cramped room with smoke and fire. As my eyes adjust, I see a man in flip-flops standing over a hole in the floor that burns bright white. He leans over with a ladle, dips it into the hole, and then lifts it to a stack of boxes that hold molds waiting for metal to become whole.

Across the room, young men open up boxes that have cooled and pull out what look like brass hair picks outfitted with pencil-thin, foot-long teeth. The teeth are snapped off and, in a corner, polished up for sale to factories that melt or shape them into products for export around the world.

Timbadia offers me a seat at his desk beneath several sketchy electrical boxes and a few dusty but still colorful images of Hindu deities. As I sit, I watch scrawny men pour sand into the boxes of unheated molds and pack them tight with their bare feet, just a step or two away from the white-hot hole in the floor. On the other side of the room is a pile of Honey just waiting to be stuffed into the fire. The air is hot, filled with soot that chokes. I feel bad breathing it; I can't imagine how it'd feel to inhale it for an eight-hour shift, much less a career.

In the course of my visit, I'm assured by some of the bigger, more established recycling industry leaders that India's environmental authorities aren't going to stand for this kind of pollution much longer. One operator, located just up the street, pointed at Timbadia's business—and smoke—and promised me that if I return to Jamnagar in 2015, "that kind of pollution will be gone." In its place, this operator assured me, Jamnagar will be home to modern brassworks.

What, then, will happen to Timbadia?

"Jamnagar is growing. There will be other opportunities."

Timbadia offers me a small, cold bottle of Thums Up cola with a straw sticking out of it, and tells me—via Sunil's translation from the

Gujarati—that he melts around 2,600 pounds of scrap per day into brass rods that are sold to small manufacturers around Jamnagar. In total, he clears around $2,000 per month, and around here, that's damn good money.

When I ask where his scrap is generated, Sunil smiles and answers: "Dubai." I'm not surprised, but—briefly—I'm struck by the disjunction between this room, and the empty streets of that modern desert metropolis. If Dubai weren't so rich, if it didn't have such a taste for imported Indian mangoes and other goods from the subcontinent, Jamnagar would be scrounging elsewhere for scrap.

As the workers behind him labor through the smoke and steam, Timbadia, in the flush of enthusiasm for our conversation, begins to tell me about expansion plans. Next month, he's traveling to Portland, Oregon, to meet his cousin, the owner of a Dunkin' Donuts franchise, in hope of starting a U.S.-based scrap business focused on exporting to India. "I think there's a good opportunity," he tells me. "Jamnagar needs more scrap."

Later, Sunil tells me that he's skeptical of Timbadia's plan, and that makes me feel better because, quite honestly, so am I. His initiative is admirable, no doubt, but even if he had the money (and the Dunkin' Donuts money certainly can't hurt), he's going to learn quickly that as much as he loves doing business with family, not even family ties can overcome this simple fact: shipping to China is going to be more profitable than shipping to Jamnagar.

As we drive through Jamnagar, Sunil takes me past the large open field that his family recently bought in the GIDC3—the third phase of the Gujarat Industrial Development Corporation's business park kingdom. He has big plans, he tells me, including a modern, technologically advanced copper rod mill to compete with the ones that his friends and competitors are already building. Coal-fired furnaces are coming to an end in Brass City, he tells me. The city is becoming too rich, and the government doesn't want them anymore.

Where, I ask, will he get all of the metal?

"I have a new trading office in Dubai," he says. "I'll source it there."

Sunil is proud of his Middle Eastern supply chain, but over the course of five days spent with him in Jamnagar, I become aware that he longs for an opportunity to buy and ship scrap from the United States, the Saudi Arabia of Scrap. Once or twice he even asks if I might introduce him to a few U.S. scrapyards that might be interested in doing business with him. I decline the opportunity, with the explanation that I wouldn't want to waste anybody's time. "Even if you guys could agree on price," I tell him, "the cost of shipping would kill a deal." He nods silently, but I can see on his stony, disappointed face that he has witnessed what I know well: the bounty of an American scrapyard.

A late morning, summer 2011, and I'm where I like to be: the Saudi Arabia of Scrap, speed-walking along bales of aluminum radiators, cables, and frames stacked four and five high in a scrapyard. David Simmons, Alpert & Alpert's burly scrapyard manager, is rushing me along. "Clean mag, irony mag, flash mag," he calls out as we weave between boxes loaded with different types of magnesium scrap. Then we're in the aluminum. "Irony aluminum, 356 aluminum, pistons."

I nod as if I know what he's pointing out. But honestly, we're moving so fast that I'm in a constant, unsettled state of having to choose between taking notes and taking a look. And there's a lot to look at: we're in the heart of Alpert & Alpert's seven-and-a-half-acre yard, home to as wide a display of scrap metal as I've ever seen anywhere. It might not be the biggest scrapyard I've ever visited (that one is in China, and by a wide margin), but it's the most diverse. If for some reason I wake up one morning and decide that I'd like to see a photo of a bunch of magnesium ramps (the kind used on loading docks, ISRI specification *Wood*), I'd call Alpert & Alpert. Likewise, if I needed to know the current price of scrap drum cymbals, these are the people who not only know that price but have several barrels ready for shipment.

What they don't have, however, is a lot of people out here. Sure, there's enough staff to receive, process, and dispatch all of this scrap. But compared to a crowded Indian or Chinese scrapyard, it's a skeleton crew. We walk between rows of barrels and bales without sight of a human. The reason, in part, is quite simply the cost of American labor: salary, insurance, and other associated costs mean that Pravinbhai Timbadia could pay his entire payroll for three months on a single Alpert & Alpert yard salary. If the cost of labor in India ever reaches even a quarter of LA's heights, that would change the flow of scrap, for sure. But with 1.2 billion people, an ever-present shortage of jobs, and breathtakingly low (by U.S. standards, at least) costs of living, that's not likely to happen any time soon.

David leads me to the large, secure shed where he stores the expensive high temperature alloys. "Hastalloy, titanium, tantalum," he drones as he points, one after the other, at boxes filled with each one.

"David?"

David swings around to a svelte, exceedingly polite, and rather regal-looking Asian male in his mid-twenties, wearing camouflage trousers and an expensive-looking knit brown sweatshirt. "Yeah," he coughs. "What's up?"

This svelte young gentleman twitches at David's brusque question, but he has too much poise and refinement to take a step back and reconsider. He's out of place here: he holds a small leather-bound notebook of the sort that may have never before in human history entered a scrapyard. His immediate purpose, insofar as David is concerned, is a question about the elements of "certain alloys."

David leans over the pricey notebook and—from where I stand—precise, tightly written script. "Okay," says David. "That's right, that's not." He spends an extra moment correcting the young man's chemistry and then leads me over to "the brasses." As he does, he mentions that the young man "is the son of a really big smelter in [Asia]. His family sent him over to learn the business."

As David and I rush along, I look back and see the young man taking a seat on an upside-down bucket, wrapping one leg over the other as he carefully writes David's instructions in his notebook. He belongs in the office, I think, not out here.

Meanwhile, David jams his hands into a box of yellow brass valves. "Probably a bastard alloy," he tells me and waves a magnet over the one that he holds. "Yep."

"Bastard alloy?" I'm not sure what he means, but he's already past me.

A few boxes down he shows me a container filled to the brim with shattered drum cymbals and then, a little farther down, a box packed with cheap-looking brass vases roughly the sizes of beer steins. My grandmother's basement used to be filled with metal just like this, swiped from my father's employees after they swiped them from my father's purchased inventory. "Indian made. We sometimes send them Honey," David announces. "They send it back to us as vases. Round and round it goes."

CHAPTER 6

The Grimy Boomtown Heat

When I was young, the family scrapyard used to buy forty or fifty automobiles per day. We'd junk them, mostly: rip out the engines, tear off the tires, and then flatten them into messy metal pancakes suitable for shipping across town to the North Star steel mill. I remember how those cars arrived at our yard: some were driven into it by their owners; a few were pulled by tow trucks; and most arrived on the backs of flatbed trucks. My grandmother would accept the legal titles at the office window, and pay out the going rate for a vehicle based upon the weight of the steel inside it.

Not all of those cars were junked. Sometimes people inexplicably dropped off a car worth keeping, and sometimes people dropped off cars with perfectly reusable tires, hubcaps, doors, chrome bumpers, and other salvageable, semivaluable parts. In those latter cases, we'd still pancake them—we were a scrap recycling yard, not an auto salvage yard, after all—but my father would allow employees to strip them for parts to take home (if he didn't allow it, they'd manage to do it anyway). After work, I'd see those parts—bumpers, steering wheels, transmissions—in back seats and pickup truck beds, bound for a Sunday-afternoon reincarnation.

The American scrap recycling industry is mostly about recycling, not reuse. But that wasn't always the case in the United States—my great-grandfather was always on the lookout for what he could shine up and resell, I'm told—and it's certainly not the case in the developing countries where so much American scrap goes. In China, in India, and across Asia and Africa, the first thing that many an importer does when opening a container of freshly delivered scrap is seek out the bits that can be fixed up and resold rather than immediately remelted. After all, a hammer is worth more as a hammer than as a hunk of steel. But in countries where waste is divided into one of two categories—one that's dropped into the recycling bin, and one that's dropped into the trash bin—that distinction is mostly lost.

So where's the reuse bin?

There is no reuse bin in your kitchen, of course, in societies where obsolescence is an obsession and many people simply can't imagine owning an iPhone 4 in an iPhone 5 world. But that doesn't mean the reuse bin doesn't exist. Rather, it just happens to be located in the developing world, where old things that once belonged to Americans, Europeans, and Japanese are refurbished and pressed into new duty on a daily basis. Poverty may inspire the practice, but pragmatism has turned it into an industry.

Taizhou, a port city of 4.6 million located 165 miles south of Shanghai, might be the beating heart of China's reuse industry. I've traveled to its waterfront in part to see where it all begins, but not much here suggests reuse of anything. Between the boats moored at the piers and the new, luxury high-rises at the concrete gates to the pier is a vast concrete wedge perhaps the width of a football field and a quarter mile long. All day long, men sweep that concrete of tiny scrap metal fragments, working their way between hundred-foot-long piles of anything metal—refrigerator doors to automobile brakes—rising fifteen feet high into the crisp sea air. The piles look like garbage, even to my biased eyes, but they are not garbage, and they are definitely not treated as garbage. Instead, in front of each is a waist-high signpost with a handwritten note placed in the frame. It includes two pieces of information: the person or business who bought the pile, and the date it arrived at the port. Nearby, there's usually a small mobile outhouse-sized shack that houses someone to guard against thieves until trucks can take the pile away for sorting.

The piles weigh 600 to 800 tons each and are Japan's top export to this growing, entrepreneurial hub of 6 million people. The pile that I'm admiring belongs to Shi Tong Qu, a thin, hardened man in his late thirties who wears big designer sunglasses, a gray polo, new blue jeans, and shiny black sneakers. It cost him around $400,000 cash, paid up front, and a trip to Japan to watch the metal loaded into a barge. He makes that trip, he tells me, four or five times per month. But it's worth the trouble: he needs a mere two weeks to cash out of a pile once it arrives in Taizhou, and he generally does so with a 10 percent profit.

Do that thirty or forty times per year, and you're bound to make some money.

I'm here as the guest of David Chiao, a Taiwanese-American scrap trader (and vice president for Atlanta-based Uni-All Group) who partners with Shi in a yard that processes metal that Chiao buys around the world (everywhere except Japan—Shi takes care of that market on his own). David is in his mid-fifties, but there's a boyish curiosity about him not uncommon to scrap men. He does his work for money, no doubt, but it's obvious that he's fascinated by how all that metal—whether it's from his home base in Atlanta, his favorite scrap scrounging sites in Scandinavia, or a wasteful Japanese scrapyard—is transformed into something new and shiny.

Earlier in the day, on the train from nearby Ningbo, he told me something that I've been mulling all morning: "I have a customer who tells me that Yiwu, a manufacturing hub outside of Taizhou, needs five hundred tons of brass every month to make pins for lighters."

"Pins?"

"Yeah, you know, the small piece of brass inside the lighter that helps it ignite?"

"Really?" I asked

"Think about how many lighters the world uses every month."

I look back as we walk away from the pier: there might be twenty piles like Shi's out there. Above them, looming, are the picture windows fronting those new luxury high-rises. I suspect that one day those homeowners will want views free of scrap metal.

Shi loads us into his BMW X5 mini-SUV and drives us down tangles of narrow roads lined by small, hutlike buildings that David characterizes as "typical Chinese living room factories." Inside might be small coal-fired furnaces where scrap is melted into simple new products. My eyes look for smoke, but instead all I see above the roofs are hills once terraced for agriculture but now overgrown with half-dead vines. The farmers are all in business now.

Downtown Taizhou doesn't have any towering skyscrapers—yet. Instead, it has traffic, and that's how I know that we're getting to the center of somewhere. Otherwise, all of the four- and five-story shopping malls, office buildings, and what might be apartment buildings look the same in the dirty morning sun. Most are covered in billboards, and most of the billboards seem to be related to manufacturing or the construction trade.

Taizhou grinds away.

It's 10:00 A.M., we turn left, and the side streets suddenly choke with traffic. "That's the reuse market," David says to me from the front passenger seat of Shi's BMW, nodding into the hazy muddle at a billboard-festooned building. "Right there."

It blends into everything else, if you ask me, but according to Shi it's the center of town. Shi drops David and me at the curb and goes in search of a parking spot. The two of us walk into the shrill, stopped traffic, the grimy boomtown heat. David leads me across the street, away from the mall, and onto a corner where two teenagers in greasy slippers stand amid power tools, screwdrivers, a pile of twisted copper wire, several frying-pan-sized steel cases obviously meant to enclose something heavy-duty, and various parts that I simply don't recognize.

As we watch, they slowly reassemble these parts—scrap parts—into an electric motor that might be attached to an irrigation pump out in a farm field, a drill press in a factory, an electrical generator behind someone's house, or a merry-go-round. It's a process that I've seen in India and other developing countries—places where smart, self-trained technicians make good livings by fixing and refurbishing what wealthier people throw away.

It's no side business, however, no niche. "Taizhou was built on reuse," David tells me. "And the government loves it. Big money." Just *how* big is the question: nobody in China or anywhere else keeps track of just how much scrap is diverted from the country's hungry scrap furnaces into reuse markets like this one. But if this market is any indication—and from the outside it's at least two square blocks of indications, complete with concessionaires selling Cokes and ice cream off the backs of bicycles—then it's assuredly significant.

Shi joins us, and we rush across the street and into what looks like an alley, but is actually an entrance into the reuse market. The corridor stretches for a city block, cluttered on both sides by neatly arranged piles of cleaned-up electric motors ranging in size from fists to barrels, with most ultimately destined to run equipment in factories located around Taizhou. But there are other uses, too: everything on the planet that runs mechanically—a fan, a cotton candy machine, a generator outside an ice-fishing house—has a noisy motor attached to it. There are millions in use right now, all over the world. Most of the ones at the reuse market were tossed away after they broke down in Japan (and the Japanese owner decided to buy new rather than repair), but Taizhou also receives millions of motors from the United States, from Europe, from Australia. What can be repaired in Taizhou ultimately makes its way here, to the reuse market; what can't be fixed is broken apart into individual metals and melted.

As we stroll, passing booth after booth, we pause beside three engineers in blue jumpsuits who drove over from a factory that makes wheels. A motor that drives one of the factory's machines burned out earlier in the day, and they're here to replace it for cheap. Shi says that the profit for the reseller will be 100 percent. Whoever threw the motor away in Japan, the United States, or Europe likely didn't profit at all.

"That's the American style," David tells me with a chuckle. "Buy new instead of reuse."

There was a time when Americans too valued used electric motors. They'd repair the ones that burned out, and only when repair was no longer an option would they toss them into the scrap heap. In fact, up until the early 1970s, American scrapyards paid workers to break open scrap motors and rip out the copper windings. From there, the copper would go to a copper or brass buyer, and the steel case would go to the steel mill. But as the price of American labor rose through the second half of the twentieth century, along with living standards, the cost of breaking motors began to exceed the value of the copper and steel extracted from them. Meanwhile, steel mills refused to melt the troublesome objects because copper is a contaminant that weakens steel; and the copper and brass makers had no interest in melting a whole object that's mostly steel. As a result, by the late 1970s the American countryside was littered with piles of electric motors (farms and farming equipment are major sources of motor scrap). Scrapyards, if they had motors, stockpiled or landfilled them. In other words, it was the worst of all possible worlds: no recycling, and no reuse.

Thus, as recently as the late 1980s, American electric motors were all but worthless—in North America. Oftentimes Chinese scrap traders could have them for free (while the American scrapyards that gave them away laughed at those allegedly foolish Chinese). Then they'd ship them to China, refurbish and reuse what the Americans didn't repair, and pay workers $50 per month to break apart the ones that couldn't be fixed. The remaining copper was sold for prices that, in those days, started around $1. Think about that: in the 1980s you could obtain, for free, something potentially worth over $10,000 the moment it hit Chinese shores. How many things are worth fifty times more in 2012 than they were in 1988? Internet and tech stocks aside, I can't think of any. Yet I guarantee you there isn't a single analyst on Wall Street with a chart tracking the Chinese price of electric motors over the last year, much less two decades.

David, Shi, and I keep walking, booth to booth. Some refurbishers specialize in little motors, some specialize in tall, narrow motors; some specialize in table-sized motors and others specialize in cleaned-up motor parts. We turn the corner into another corridor, also full of motors and—halfway down—a booth that sells stacks and stacks of refurbished gears as small as tea saucers and as big as large pizzas. Opposite them, in the same booth, are chains—large, industrial, bikelike chains—that connect motors to whatever it is that they're supposed to turn.

Suddenly I realize that here, in the corridors that sell things that drive other things—motors, gears, chains—the salespeople are women. *All* of them. They sit on stools, knitting, gossiping across the aisles, barefoot and—it seems to me—bored and occasionally a little bitter. "Their husbands are all out making sales," David tells me. "You can't just run this business from the market. So the men are out, and the women watch the shop."

"They're family businesses," I say.

"Exactly."

We turn another corner, and suddenly we're in a heaving space perhaps the size of two square city blocks. Bicyclists ride through raucous, crowded aisles stocked with refurbished goods ranging from wrenches to telephone cables, as if they're traversing city streets. Women stand with arms crossed, waiting for customers; old men sit in deck chairs and watch the customers pass. Above, sunlight plummets through a glass roof and sheets of plastic heavy with dirty rainwater, lending the place a dark, aquarium-like shimmer.

I walk slowly, past booths where hundreds of used drill bits—many two and three feet long—are stood on end. "In Japan they use them a few times and then send them out as scrap," David says. "They don't even need to refurbish them here. They just resell them." Then the electrical products start. There are refurbished fuse boxes, power cords, and power strips. Nearby, the refurbished power drills occupy a long aisle. Most are Japanese, with Japanese voltage requirements. But it's no matter: the guys selling them can retrofit the drills for Chinese voltage by switching out a couple of parts and a new cord.

It goes on and on. David stops and buys me a Coke from a vendor pushing a cold drinks cart through the market. Then we pass vendors selling refurbished wheels of the sort that you'd find on dollies and grocery carts. They're a colorful sight—red, blue, yellow, green—and they're all refurbished, pulled from loads of stuff that the Japanese can't be bothered to reuse or fix on their own.

The reuse market opens at 4:00 A.M. "The factories get started early around here," David reminds me. "And many just run all night. If you need a motor at four A.M., this is where you'll find it."

Taizhou's reuse market is one of China's biggest, but it has counterparts in every city, town, and village across China. Sometimes that market is big, like Taizhou's; sometimes it's just a row of refurbished motors and televisions in front of somebody's home. Not everything is imported; more and more, China refurbishes what China makes and wastes. But it's everywhere: reuse is integrated into the Chinese economy as tightly as annual car models, regularly updated iPads, and the latest laser disc movie format are built into the American one. That won't be the case forever—and these days, it's less the case than it was twenty years ago—but compared to the economies China seeks to emulate, it's still way ahead in deploying the reuse bin.

Here's what may not be well appreciated in countries that fetishize the new and upgraded: Taizhou may embrace the reuse of what others throw away, but it's far from poor. According to Shi, it enjoys the highest per capita car ownership in all of China, a status in no small part owed to Geely, a native Chinese carmaker with so much success it managed to buy Volvo, outright, in 2010. Geely is the biggest, but by no means Taizhou's only manufacturer of note. Some of China's best-known bicycle and motorcycle manufacturers are here; so are makers of appliances ranging from washing machines to air conditioners.

All of this manufacturing—whether of cars or reinforcing steel for new shopping malls—is fed to a considerable degree by imported scrap metal. But unlike Guangdong, where much of the metal is then reexported as new goods to the countries that sent it, Taizhou's imported scrap mostly stays in China. The cars are sold in Taizhou's auto dealerships; the refrigerators are marketed in Shanghai shopping malls; the bicycles are sold in the dusty towns of west China. Without that imported metal, and the demand for that metal all over China, this town would be just another stagnant backwater wishing it had raw materials of its own. Without it, there'd be no Geely, no auto parts manufacturers, no air-conditioner makers. There'd just be farms.

This hunger for scrap is, in part, a hunger for the chance to develop into middle-class consumers. It makes no difference whether it's achieved with American scrap, Japanese scrap, or European scrap. The important thing is that Taizhou has scrap—especially motor scrap—in endlessly arriving containers. This hunger, however, has started to have a very real, very troubling effect (if you're in the motor scrap export business): there are fewer electric motors to import from abroad. Over breakfast David told me, "In the early eighties, I saw shiploads of motors, from the Chicago area, the Great Lakes area, and [the barges would be loaded with more] going down the Mississippi River all the way to New Orleans. They'd keep on feeding in St. Louis, they'd keep on feeding in Memphis, and all the way to New Orleans. And in New Orleans they feed into the bulk ship, like twenty thousand metric tons. It's not there anymore."

It's not there anymore for two reasons. First, the piles of motors that used to litter the American countryside in the 1980s and 1990s have already been exported. Now, the scrap motor market is limited to what's being thrown away in real time in the United States, Japan, Europe, and—increasingly—China.

The second reason is tougher, if you're looking at it from the perspective of an American scrap man like David. It goes like this: the United States simply doesn't have as many factories as it did in the 1980s, and so the United States isn't wearing out as many factory motors as it did during American manufacturing's boom times (motors from factory equipment were a major contributor to the U.S. motor stockpile). Some of that's due to efficiency, and some of it's due to global manufacturing's ever-creeping migration to Asia. But if you're an American, the results should give you pause: the motors that used to drive U.S. industry are being exported to China, refurbished, and used to drive Chinese industry. What can't be refurbished is a leading source of copper for washing machines, air conditioners, and other appliances and luxuries coveted by China's growing middle class.

As Shi drives us out of Taizhou, he talks about skyrocketing real estate prices, the availability of luxury cars, and the future luxury mall scheduled to open soon in downtown, not far from the reuse market. Meanwhile, the road opens, then narrows, and suddenly we're surrounded by trucks, motorcycles, bicycles, and men loaded down with scrap cables, wires, sheet metal, and rubber insulation. This is Taizhou's recycling park, a government-designated zone where thirty-four of the city's biggest recyclers (out of more than a thousand, most of which are small family-run operations) are corralled so that they can be more effectively regulated (that's the theory, at least). From Shi's back seat, however, it's gloriously chaotic—a place where metal of any possible variety, shape, and use moves down the streets on bicycles, backs, trucks, and automobile rickshaws. We pass open factory gates, and inside I see men and women stripping wire, sorting shredder residue, and loading clean metal into trucks.

Shi's company is located in a shabby two-story brick building on a main boulevard. When I step into the murky marble lobby I don't get the impression that there's much office work associated with this kind of company. Two broken wicker chairs are shunted off in a corner, the plants are wilted and dusty, and the floor looks as if it's survived a rocket attack.

We walk outside and through a gate into a long roadway that runs for perhaps a thousand feet between two rows of ugly, mixed-up scrap metal, interrupted here and there by pockets of workers. It's the same hellacious dark metal Japanese jumble we saw at the port, but here the jumble begins to make sense as workers sort through and separate it into (relatively) clean components. On my right are piles of scrap water meters amid the sheet metal; to my left cables are twisted among piles of what looks like shredded aluminum siding torn from a house. Nearby, twenty shovel-like rubber pans are lined up in a square formation. Some are filled with wire, others contain just a few pieces of copper, and still others hold fragments of electrical components that I can't distinguish. Yet they obviously mean something to sorters trained to recognize value by the fragment.

David explains to me that Japanese scrapyards are so lacking for space—especially in Tokyo—that they just toss their scrap into big piles, load it into containers for shipment to Taizhou, and await buyers like Shi. What I see here is the result of their waste—a sort of scrap-metal mystery grab bag. No other developed country on the planet tosses its scrap metal away like the Japanese. But the Japanese simply can't be bothered, in large part because they know Taizhou will sort out the problem for them. Nobody around here is complaining, though. Shi, for one, has grown rich off that Japanese grab bag.

There must be fifty sorters here, but it's hard to tell: new ones appear with every step I take, hidden behind piles of scrap, their gloved hands sifting through wild mangles of metal. I lean down to get a better look at a cluster of five workers sorting through the smallest fragments: brackets, screws, electrical connectors, circuit boards, brass rings of no particular provenance, aluminum crescents of even less provenance, sharp gears, fragments of chains, bits of wire, broken pipes, cans, bits of radiators, and fingernail-sized flecks of something. Each piece, on its own, is nothing; each bucket is little more than nothing; but weeks and days of so much nothing can add up to millions of dollars, and the raw materials that have raised up Taizhou.

We continue deeper into the complex and stop beside two muscular women working with hammers and chisels to break apart bucket-sized electric motors that once drove machines in Japanese factories. Breaking up motors is an art form, with the best, most efficient motor breakers earning as much as $500 per month to dismantle something that most people don't realize even exists. The case must be broken into—no easy task—and then the copper windings torn away while the other parts are segregated. Anybody can do this, of course, but it takes nimble fingers, strength, and experience to do it quickly.

Nonetheless, the greatest efficiency is embodied in the motors pulled from all of this wreckage and set aside for delivery and sale to those who fix them up for reuse. It's the treasure in what looks like trash, a profit beyond recycling.

It's the reuse bin.

Shi tells me he imports more than $1 million worth of electric motor scrap per month in addition to all that Japanese scrap, and he'd import more if he could. Most of it is scrapped, but if some reuse can be pulled from the piles, all the better. "Taizhou needs the metal," he says several times over the course of the day I spend with him. It occurs to me that Taizhou certainly needs the metal more than the Americans, the Japanese, and the Europeans, who sent it here. Back home, there wouldn't be a reuse pile; there'd just be a stockpile waiting for someone with the good sense to see some value in what others throw away.

Take, for example, that old computer monitor in your closet. In the United States it's a bunch of raw materials waiting to be turned into something new, but in Africa, in developing Asia, and in parts of South America it's something much better: a low-cost means to get someone onto the Internet. All that's needed to make it happen is a shipping container, somebody with the technical skill to fix it, and someone to sell the new monitor to someone who wants it.

These sorts of businesses are everywhere.

Days before Chinese New Year, 2011, and I'm in Penang, Malaysia, wandering through a small shop where used computer monitors are stacked atop display cases. The price is cheap: I could get something nice and big for $50. If I want a computer to go with it, those are available too, for a touch more money. There aren't any other customers at the moment, but that's expected: I'm standing in the factory outlet of Net Peripheral, one of Malaysia's best and biggest computer monitor refurbishing companies. If I'd like to see their products in a more flattering, crowded venue, I just need to visit some of the shops they supply.

The company's managing director and cofounder, Su Fung Ow Yong, a petite and tough early-middle-aged woman, walks me out the door behind the shop and into the company's warehouse. There I see a solid brick of computer monitors stacked five high, end-to-end, and three deep. They just arrived from the United States, bound together by cellophane, and a forklift is moving them into place among dozens of similar stacks. There are, best as I can tell, thousands of used, imported monitors here, many bought from a Vermont-based company that's shipped well over 300,000 monitors to just this factory over the last four years.

Here's the thing: the monitors that Fung imports are not "dumped," and they're not even recycled. Rather, they're deposited into that elusive third bin: the reuse bin.

Who wants an old computer monitor?

People who don't earn enough money to buy new laptops, desktops, or smartphones. In other words: most of humanity (for example, less than 5 percent of India has access to a computer). Exporters who directly supply used computer equipment to the Middle East estimate that more than half of all computer purchases in some parts of Egypt are secondhand equipment. Likewise, a recent European Union–sponsored report on Ghana's imports of used electronics determined that 70 percent were in fact destined for the reuse and refurbishment market (with 15 percent destined for recycling). In other words, Egypt's "Twitter Revolution" of 2011 didn't take place on new iPhones (which most Egyptians can't afford), but rather on repaired five- to ten-year-old desktop computers and monitors that were exported from the United States, the European Union, Japan, and elsewhere in the developed world. Put differently: if used monitors and computers didn't leak out of American closets, there would have been fewer people online in Egypt when being online made a difference.

Companies like Net Peripheral were developed to supply markets like Egypt's. They aren't alone: refurbishment exists wherever people can't afford new (it also exists, on a lesser but no less committed scale, in developed countries among people who appreciate the ethical and thrifty value of reuse). From the 1990s into the mid-2000s, China was home to a huge, highly profitable monitor reuse and refurbishment industry. These days, India, Mexico, and Africa are home to thriving, and growing, reuse industries as well. The potential profits are huge. During the glory days of the Chinese reuse and refurbishment industry, monitors were bought in the United States and European Union for less than $10, refurbished in China, and then resold for $100. The trade was, however, a victim of its own success and of the changing times: the Chinese government shut down these businesses to protect new monitor manufacturers, many of whom are government owned or connected. Meanwhile, China's consumers got rich and started coveting flat-screens.

At Net Peripheral, refurbishment starts not long after a monitor arrives at the facility—if there's a scuff on the case, workers polish it out. But refurbishment isn't just a matter of touch-ups: a nearby table is covered with monitors plugged to a power supply that allows Net Peripheral employees to judge whether or not there are blurring problems (the most common issue upon arrival—and 100 percent repairable). Farther along, I find two Indonesian technicians busy opening monitor cases and methodically clipping, poking, and soldering the electronics on the inside, repairing and replacing electronics that failed—for whatever reason—in the United States. They work like surgeons, removing and replacing parts—some ordered from factories in Taiwan, and some recovered from unrepairable monitors that were broken down for recycling—until the repaired monitor's innards work as well as they did when it was new.

From there, the monitor's innards move down a conveyor to technicians who test them to make sure they broadcast a good picture. Next, they're placed into new or refurbished cases, packed for shipping, and sent to customers worldwide. It's a profitable business, and it employs sixty workers who receive good, competitive wages for their work. But in the end, even Fung has to concede that its future is limited. At the time of my visit, she and her husband were thinking actively about relocating from Malaysia, where living standards had risen enough to interest people in buying new monitors, to Indonesia or a different developing market where used monitors might have a customer base.

It's a bitter irony: the wealthier Malaysia becomes, the less likely it is to embrace the thrifty practices of its developing stages. Later, as Fung drives me over the bridge that connects Penang to peninsular Malaysia, she nods at all of the new cars. Malaysia is home to a burgeoning domestic auto industry, and its roads are becoming as crowded as Manhattan's at rush hour. "If you can buy a new car, why would you buy an old monitor?"

It's a good question. As Fung drives, we pass some of Penang's high-tech industrial core: factories owned by Dell, Sony, and Intel. Who wouldn't prefer what emerges fresh from their loading docks? On a cloverleaf I look down and see piles of rusty scrap metal spread out for acres beneath cranes and tractors. Those rusty hills remind me of home, of the scrapyard in which I grew up. The scrap recycling industry is growing in Malaysia, just as it grows anywhere people enter the middle class and leave behind their thriftier ways. But Fung, for one, won't likely hang around Penang to see it happen. She's looking for somewhere new, somewhere there's a demand for the reuse bin.

CHAPTER 7

Big Waste Country

Monday morning, just before 8:00 A.M., Johnson Zeng eases his rented Chevrolet into a space in front of Cash's Scrap Metal & Iron in St. Louis, Missouri. He's in the market to buy scrap metal that he can ship to China, and this is the first stop of the day in the middle of a two-and-a-half-week road trip to regular customers that started in Albuquerque, New Mexico, and will end in South Carolina. But that, Johnson assures me, is nothing. "My last trip with Homer," he recalls, referring to the scrap importer in Guangdong Province who provides him with most of his business, "we drove ninety-six hundred miles in twenty-six days."

The result? Millions of pounds of metal worth millions of dollars left the United States for China.

Johnson is one trader, in one rental car, traveling across the United States, in search of metal. But he's not alone: by his estimate, there are at least one hundred other Chinese scrap-metal traders, more or less like him, driving from scrapyard to scrapyard, *right now*, in search of what Americans won't or can't be bothered to recycle (there are paper traders, too, but in much smaller numbers). It's an old trade—Taiwanese buyers like Tung Tai's Joe Chen did it in the 1970s—and it's essential. He and his peers are the eyes that spot the value that Americans refuse to see; they're at the vanguard of sustainability, the high-tech scavengers of an endlessly upgradeable age, the greenest recyclers in an era when that means something. In a very real sense, Johnson Zeng is the link that binds your recycling bin, and your local junkyard, to China.

Today Johnson is on North Broadway Avenue, not far from the Mississippi, in an industrial neighborhood that's seen better days. Truck trailers fill empty lots, and the sidewalks are vacant and dusty. Those lots, I imagine, must have once been occupied by warehouses and factories. Now the only thing you can say about this place is that it's not somewhere you want to visit after dark.

Johnson clicks away at his BlackBerry, checking the London metal prices. "Market is down." He sighs. "But we will still try." He's a young-looking forty-two, but when his lips purse with concern—as they do now—his cheeks puff out slightly, highlighting the lines at the corners of his eyes, aging him. He has a high forehead that gives him a justifiably thoughtful appearance, and a soft voice that, combined with his polite English, suggests a gentle, refined character. "Homer calling," he says with a whisper as he presses answer, and those soft-spoken English words are bulldozed by the throaty, aggressive inflections of his native Cantonese. I've quickly learned that Johnson, native of Shantou, in northern Guangdong Province, likes calls from home.

Today, like most days, Johnson and Homer are interested in copper, and for very good reason: in 2012, China accounted for 43.1 percent of all global copper demand, or more than five times the amount demanded by the United States that same year. Why? One reason is that China is growing fast, and a modern economy can't grow fast without copper. But the other major reason is that the last of the American factories devoted to refining copper from scrap metals shut down in 2000 due to the high cost of complying with environmental regulations (and in part due to enforcement actions against those who wouldn't—or didn't—comply). Partly as a result, China, which barely had a copper refining business in 1980, now has the world's biggest. Not only that, it has some of the best, most technologically advanced (and environmentally secure) copper refineries in the world. So when Johnson goes out on the road to buy copper, he's buying copper that might have once stayed in the United States, but now has nowhere else to go *but* China.

That material is generally known as low-grade in the scrap-metal industry. It's a vague but important term that means different things to different people. But, in general, low-grade scrap requires significant work—manual, chemical, or mechanical—to turn it into something high-grade, along the lines of the chopped-up copper wire and cable that I saw at OmniSource in Indiana. For Americans who care about recycling and preserving resources, here's the most important thing to know about low-grade scrap: most likely, it'd end up in a landfill if it weren't exported. It's just too expensive to recycle without cheap labor to extract the metal. Electric motors, like those valued in Taizhou, are a premier example; but so are Christmas tree lights and any kind of cable packed in lots of insulation.

As low-grade scrap buyers go, Johnson is midsize at best. But midsize isn't modest: the previous night he told me that he'll try to spend $1 million before the end of the week.

Johnson wraps up his phone conversation with Homer and slips the BlackBerry into his front shirt pocket. "He's waiting by his computer," he tells me as he grabs the car door handle. "I'll send him photos."

I check my watch. It's just before 10:00 P.M. in China. "He's staying up?"

"Of course! Some of the [scrap] material, I don't know what it is. Only he knows. So I call him. He's the expert." Johnson steps out of the car and opens the trunk. Inside is his suitcase, my suitcase, and a hard hat. He opens his suitcase, takes out an orange safety vest of the kind worn by highway construction workers, and slips it over his freshly ironed blue-and-white-checked shirt. Then he reaches into his wallet and takes out a business card that he slips into the clear plastic holder sewn into the vest.

JOHNSON ZENG
PRESIDENT
SUNRISE METAL RECYCLING
VANCOUVER, B.C.

He straightens his back to his full height, perhaps five-nine, and brushes down the wrinkles in his vest. At that, he shuts the trunk and we walk through the front door of Cash's Scrap Metal & Iron. There's a window on the other side, a slot through which documents and money can be exchanged, and—on a rickety chair—a sleepy man in a hard hat and greasy clothes who tries to avoid my stare.

"Hello?" Johnson says through the cash slot.

A meaty middle-aged female face appears in the window, laughing, apparently midconversation. "Can I help you?"

Johnson stands straighter, smiles broadly, and slips a card through the slot. "Good morning, ma'am!" He drags out each syllable with a fawning inflection. "I'm Johnson with Sunrise! I have an appointment with Michael [not his real name]!"

I look at him: Where the hell did the cold-blooded Cantonese businessman go? This is not the guy I flew to St. Louis to meet.

The woman looks at the card. "He's not in."

I see Johnson flinch. "No problem, ma'am! Do you know when he'll be in?"

"Lemme check." She walks away from the window.

His smile drops. "Always like this," he whispers. "Always."

I hear a phone ring on the other side of the glass; a diesel motor groans on the street.

The door opens just slightly to reveal a tall, muscular man in his early thirties wearing a red T-shirt. His hand remains on the knob, and his body leans in the direction of the place from which he came. "Hi, Johnson, I'm in the middle of payroll." He nods at a ratty leather sofa in the middle of a bedroom-sized office. "I'll be with you as soon as I can."

As he walks off, Johnson flashes a broad, toothy smile. "Take your time! No problem!"

We sit down, and I take a careful look at Johnson. I can't imagine wanting anything—even scrap—so bad that I'd transform my public persona so completely for a chance at it.

"I made an appointment last week," he says, his normal soft inflection returning with a touch of bitterness. "It's always like this. Always."

I look down at the worn and uneven linoleum floor, permanently stained with what looks like decades of grease and dirt. The harsh fluorescent light above us highlights every last capillary crack, every corner torn away, every ancient spill. Still, this dinginess is nothing new to me: my family's scrapyard wasn't much cleaner, and—cold reception aside—I feel right at home. My grandmother and I spent some of our best hours in places like this. Maybe somebody here feels the same way, but there's nothing here—a child's drawing, a family photo, a coffee cup with a grandchild's face on it—to suggest that any of this means anything to anyone.

I look up. The receptionist chats on the phone about the weekend and—briefly—scratches her rear end. She doesn't look at Johnson or me. We might as well be cushions on the sofa. I look at the calendar on the wall: one month out of date.

"What're we looking for today?"

Johnson and I look up from the sofa and the man in the red T-shirt is standing above us, wearing a hard hat and holding a clipboard.

"ICW," Johnson replies, using the universal shorthand for insulated copper wire. "Also, radiator ends." Red T-Shirt hands me a hard hat, and we follow him out a door at the back of the office and into a narrow, cramped warehouse lined with dozens if not hundreds of washing-machine-size cartons stuffed with scrap metal of various types.

The light is dim, and mostly comes from the sunlight streaming through the loading docks. Red T-Shirt knows we're behind him, but he walks too quickly, as if he's in a hurry. Still, Johnson goes at his own pace, his eyes moving up and down the various piles of scrap. His public manner may be fawning, but now—amid the scrap—he has become forthright, serious, and intensely focused on the random assortment around him. Red T-Shirt points at a carton of cables wrapped in dirty canvas. "Lot of this just came in."

Johnson takes the BlackBerry from his breast pocket, holds it over the box, and snaps a photo. "Elevator wire," he says, and double-checks the image before pressing send. "Goes to Homer." He moves to the next box—it contains a mix of wires of various types and colors. Some are thick; some are thin; some have small metal connectors; and some are frayed and showing the thin copper strands inside.

In the United States and Europe, this mix of wire is classified as ICW—insulated copper wire—and sold at a single price. But once it goes to China, the red wire will be separated from the green (there are different percentages of copper in the different types), thick from thin, and with connectors from without. Each type will have its own price and—often—its own market. For Johnson, that's a matter of paying, say, $1 per pound for something that contains products priced at $0.60, $0.80, $1.20, and $2.20 per pound in China. But as much as Johnson knows, Homer knows more about China's local markets and what they'll pay for the various types. For the most part, non-Chinese don't know about these micromarkets, and even if they did, they wouldn't be able to access them due to language and culture. So Johnson takes a photo and sends it to Homer. "How much do you have?" he asks.

Red T-Shirt looks at his clipboard. "About eight thousand pounds. How about jelly? We've got probably ten thousand pounds of that." He points to a box of two-inch-thick cables cut into one-foot segments that ooze a Vaseline-like substance and hundreds of tiny wires. In a past life those small wires transmitted telephone conversations underground, and the "jelly"—a petroleum product—repelled the underground moisture that threatened to corrode them. American wire recyclers don't like it because the jelly gums up the blades in their chopping equipment, so they ship it to China, where it's cut apart by hand and washed clean with soap.

Low-grade, indeed.

Johnson snaps a photo and transmits it to Homer. Then he notices something: "Ah, Christmas tree lights."

The lights are loose in a box, and Johnson reaches to pull away at the knots, hoping to see if there's anything beneath them. "Not good quality," he whispers to me. "They should be baled." That is to say, they should be compressed into a cube so that nothing can be hidden beneath them in a box. Johnson looks over to Red T-Shirt. "Maybe should be priced less."

"Nah, Christmas tree lights are Christmas tree lights," Red T-Shirt responds.

Johnson stares at the box and clucks his tongue. He wants them.

Red T-Shirt moves along. "Our ACR ends are down here."

Johnson walks over to a box that holds twelve-inch strips of metal that look as if they've been knitted with single strands of copper tubing. Those strips are made of aluminum, and the tubes once circulated radiator fluid. The radiators are elsewhere, likely sold to an aluminum remelter; these are the aluminum-copper radiator (ACR) ends. They're perfect for a cheap labor market like China, where people can be paid to clip the copper loops from the aluminum.

"How much do you have?" Johnson asks as he takes a photo.

"Around ten thousand pounds, I think."

This is how it goes for the next ten minutes: Red T-Shirt shows Johnson cable television cables (not interested), cable television boxes (very interested), power lines (very, very interested), and the other essential accessories of daily life that eventually get thrown away, but are never counted as household waste. Johnson takes photos of everything, and carefully writes down available volumes.

"Do we have enough for a container?" Red T-Shirt asks.

It's a key question. A standard forty-foot overseas shipping container of the sort that Johnson will send to Homer in southern China can hold 40,000 pounds. But here's the catch: the high cost of moving containers from one scrapyard to another means that a container can only be filled up in one scrapyard, and that means Johnson needs to buy 40,000 pounds of scrap metal at Cash's—or nothing at all. It makes for a tricky transaction because even if Johnson finds great buys on radiator ends and insulated copper wire, he can't close on them unless they add up to 40,000 pounds. Thus, he may have to take a loss on some material to make a profit on what he really wants. He runs his finger down his notes and purses his lips. "We still need 10,000 pounds. How about some Christmas tree lights? Do you want to sell them?"

"Let's go inside, and I'll check how much we have."

We follow Red T-Shirt back into the office and take a seat on the ratty sofa. But there's no time to relax: Homer's number is flashing on Johnson's BlackBerry. "Must have some new prices," he says and answers. The conversation lasts less than ten seconds. "Homer is cautious today," he tells me as he hangs up. "Market is down. But we will try."

At that he opens his notebook and takes out a sheet of paper marked "Purchase Order." It's a crude form, obviously made at home on a PC, and—in addition to Johnson's name and company—it contains the three columns that matter: material, weight, and price. Slowly, he writes:

Jelly Wire with steel	10,000 lb	55
Grease Wire	5000 lb.	135
#2 Ins. Copper Wire	8,000 lb	150

He's beginning the fourth item when he receives another call from Homer. It's a ten-second Cantonese machine gun burst. Whatever was said, it's enough for him to cross out the first jelly wire price and raise it to 56 cents. He writes down an additional seven categories, and by the time he's done he's offered to buy close to $60,000 worth of old wire and hardware. "Maybe not competitive enough today," he worries. "Let's see."

Red T-Shirt looks around the corner. "Stu will be with you in a moment, Johnson."

Johnson nods to himself. "We used to be able to buy five to eight containers at a time from this yard," he tells me. "Now lucky to get one. More competition. Some days the yards have two or three teams of [Chinese] buyers come. Seller's market."

"Saudi Arabia of Scrap," I answer.

"Maybe." He nods. "Maybe." He places his hands on his knees, takes a deep breath, and then pulls up the current London prices on his Black-Berry. For a moment, it's quiet but for the roar of a machine on the other side of the walls.

I spend six days on the road with Johnson, scrounging for containers of scrap metal worth as much as $100,000 each, eating off-menu at Chinese restaurants, and sleeping in Red Roof Inns. Some days we'd drive six hours only to find that the scrap Johnson had been promised was sold a few hours earlier to another roving Chinese buyer; other days, Johnson spent the equivalent of a Lamborghini on scrap. Scrapyards, however, were usually just the intermissions between long, introspective talks as we rode cross-country.

Johnson tells me that all of the time on the road can be lonely and frustrating—especially if you've been doing it for five years. He has a wife and son in Vancouver, but he only spends about six months of the year with them. The rest of the time is spent driving across the United States in rental cars, buying scrap, and thinking. "The last couple of years I spent too much time thinking about life, work, family," he tells me. "I even began to think of being a Christian." What finally eased the pain was not religion but reading Eckhart Tolle's book *The Power of Now* and observing Homer's powerful example on the road, and in motels, across America. "He's very filial," Johnson says. "The first thing he does when he gets to the hotel room is call his mother. I learned a lot from him. I learned to stop worrying about the things I can't worry about."

Nonetheless, there's one thing that he continues to worry about on a daily, sometimes hourly, basis: where to eat. During his early years on the road, it was a problem solved by thumbing through phone books. These days, his GPS is filled with directions to the best Chinese restaurants across the United States. When we stop in, many of the owners recall him, and even more recall Homer, who, I'm told, is a charmer.

But even after so many years there are occasional gaps in Johnson's GPS.

One night, for example, while driving through rural West Virginia, we were both hungry, but Johnson didn't have anything in his GPS. I eyed signs for fast-food stops.

"Hooters!" Johnson exclaimed.

Ahead, at the top of an exit ramp, a neon version of the distinctive logo was lighting up the Appalachian night. "Really?"

"Sure!" He rolled up the exit, all the while explaining to me that his American customers often take him there for lunch, and he really likes the spicy chicken wings.

"You know," I told him delicately, as we took a seat in the midst of the restaurant, "they have Hooters in China, now."

He glanced at our waitress in her tight tank top and hiked-up running shorts. "And they dress like that?"

"They do."

"Hmm."

This wasn't the future that Johnson had foreseen in Shantou, a modest city in northeastern Guangdong Province. His father was a well-known agricultural scientist of whom Johnson speaks with pride and devotion; his beloved mother was a farmer. In school Johnson was a good student who excelled in the sciences and earned a degree in polymer science—plastics, basically—in 1991. Back then, a young college graduate didn't need to look for a job; the government gave them out. Johnson landed at a plastic plant owned by SinoPec, a state-owned oil company. "I started as a worker, then went to supervisor to vice department head to vice general manager," he tells me. "I went from the marketing department to the finance department. The factory had five hundred employees."

In 2001 he was promoted again and given the opportunity to set up another factory worth—in today's terms—about $12 million. "So why did I move to Canada?" he asks. "I didn't want to leave. I liked the achievement feeling. It's very important to me, even better than money.

I had my apartment, my car, a good salary. Nobody believed I would leave. 'You're young, with a good future.'" However, Johnson's wife, who also worked at the chemical factory, had grown tired of the odors, the work, China. Meanwhile many of her friends were emigrating to Canada and telling her that life in the West was better. "So I said, 'Let's do this. Let's change.'"

That was the easy part.

Johnson went from being a young man on the move with a guaranteed future in a Chinese state-owned company, to working odd jobs in Vancouver. He started out as a renovation contractor, moved on to selling fruit in Chinatown, and spent several years working in the dairy department of a supermarket. Then one morning in 2006, he was reading a Chinese-language newspaper and came across an advertisement for a "trader" that specified neither the company nor the goods traded. "This sounds good to me," he recalls thinking. "Trader! It's my favorite job! Marketing!" A few weeks later he sat down for an interview with an official from a Chinese scrap-metal-buying cooperative based in Vancouver. They were looking for someone to join their sales teams prowling North America for scrap metal, and they offered him $1,200 Canadian for three months' work, plus a $300 bonus if the work took him to the United States. "Why did I take this when I could earn more money for my family in the supermarket?" he asks me. "Because I started thinking about plastic recycling back in China years ago. So I thought this was a way to learn and start."

His new employer gave him a week's worth of training and sent him on the road in a rented car. "I was supposed to buy things I didn't even know the English name for!" No surprise, the first week was a disaster: he bought nothing. But Johnson is smart, cagey, and brave, and over the next three weeks he managed to buy thirty-one containers of scrap for his new employer—a take worth well in excess of $750,000. By 2008 he knew enough, and he was successful enough, that he could contemplate leaving and setting out on his own. At exactly the same time, Homer, a member of the cooperative that had hired him in the first place, was also contemplating a business of his own, and unlike Johnson, he had a scrapyard in Guangdong to which Johnson's purchases could be sent. "After I went on my own with Homer," he tells me while driving through Kentucky, "I spent seven months on the road. One time I spent seven weeks on the road with Homer without going home once."

"How much scrap did you guys buy?"

He thinks for a moment. "Hundreds and hundreds of containers."

Millions and millions of dollars.

At Cash's Scrap Metal & Iron in St. Louis, we wait five minutes, then ten minutes, but nobody comes to take Johnson's purchase order. While we wait, he stares at the London prices. Then he looks at the Chicago prices. He asks if I mind Chinese food for lunch, and of course I tell him I don't.

"Johnson?" bellows a big voice. "C'mon in!"

Johnson rises from the sofa and strides into a corner office that overlooks a large, messy desk where Stu Block, the portly, curly-haired founder of Cash's, leans back in an office chair like a scrap-metal potentate. There are three other men in the room, and they smile as if they've just been privy to a particularly dirty joke that they've promised not to share. It's a testosterone-fueled, somewhat ironic atmosphere that Johnson totally shatters with an enthusiastic: "Hello, sir! How are you today?"

"Fine, Johnson." Block's eyes turn to me. "Who's your friend?"

Johnson introduces me as a journalist following him to learn about the life of the Chinese scrap trader. At the word *journalist*, Block lights up and tells me about the day that Mike Rowe and *Dirty Jobs* spent filming a segment at the plant. Rowe's work shirt, he tells me, is even framed, signed, and hanging in the office.

"I'll take a look at it," I tell him.

"You should!" He turns to Johnson. "Anyway, what you got today, Johnson?"

Johnson hands over his purchase order, and Block runs his eyes over it with a crooked smile. "All right. Let me think. I need to see where the market's going. I'll have somebody call you later."

I turn to Johnson. I look at Block. That's how he responds to a $60,000 purchase order for something that can't be sold to anybody in the United States? Who the hell else is going to take those Christmas tree lights?

"Thank you, sir!" Johnson says. "I'll call later."

"Take care, Johnson."

We walk out the door, and as soon as we're on the sidewalk I burst out: "He barely looked at your prices! Jesus Christ. He's not interested?"

"Probably not. Too many other buyers. I can tell other buyers were here yesterday. There wasn't as much scrap as usual." He opens the car and places his hard hat and vest on the back seat. "It's okay. Tomorrow we will arrive somewhere before they do."

As we settle into the car, he reaches into the glove box for the GPS. It contains the names of dozens of scrapyards—Johnson's client list—and he punches up the next one we're due to visit.

"I was surprised to see so many Christmas lights," I tell him.

"Big waste country, the U.S.," he answers. "They make these things but have no way to recycle them. Not enough copper in Christmas [tree] lights for the big companies . . . to chop them up and be worth the money. So we buy it."

Without a map, I have no idea where we are in St. Louis. It could be anywhere, but Johnson doesn't seem disoriented. The GPS is his English-language muse: it says turn left, and Johnson responds softly, mostly to himself, "Oh really?" He's quiet for a moment, sighs, and then glances at me. "So you can see how hard the life of the Chinese scrap trader is!"

Johnson is a good driver. He follows the instructions of the GPS without question, always signals his turns, and rigorously follows the speed limits. As we approach downtown St. Louis, my eyes are distracted by Busch Stadium, home of the St. Louis Cardinals baseball team, and the half loop of the Gateway Arch, which, just then, reminds me of the copper tube on a radiator end. When I turn to Johnson, however, his eyes are on the road. "Have you ever been up there?" I ask, pointing at the arch.

"Noooo. I've been here twenty or thirty times but never go up there."

"What do you do after work?"

"If I'm with Homer, we have dinner in Chinese restaurants and then go back to the hotel. Then he spends all of his free time chatting online with his family and watching Chinese television shows. Like he's in China. His wife makes dried sweet potato for him that he eats during the trip. One time we went to McDonald's and he was sick for three days. So now only Chinese food. French fries are okay, though."

"Really?"

"He used to come out with me more. But when I got the BlackBerry he could stay home with his family and just wait for the photos."

Appropriately, Johnson's phone buzzes with Homer's number. "Maybe he has some new prices for me. Let's see." But when he answers, the connection breaks off. This brings to Johnson's mind something else. "We travel so far together. After every trip I kiss the car." He pats the steering wheel. "Thank you for the safe trip."

Homer rings again, and this time the call connects. Johnson's voice goes deep, and he's back in Guangdong.

According to Johnson, Homer spends his mornings drinking tea with the customers for his metal. They gossip about friends, family, and the economy—and thus by lunchtime Homer knows what kind of metal he wants Johnson to buy, and what kind of price he'd like to pay for it. There's no fancy trading floor in Homer's hometown, just factories and tea-drinking sessions. It's all very simple, and it's a very big part of how the market in America's old phone cables, Christmas tree lights, and other things that get shoved into the backs of closets and garages is set every day.

At 6:30 A.M., Tuesday, Johnson and I pull out of Louisville, headed for Indianapolis. It's a two-hour drive, and to save time Johnson pulls off the interstate into a Wendy's drive-through for breakfast. It pains him: he prefers to eat the free motel breakfasts he finds at Super 8s, Red Roof Inns, and other budget chains. But we're running late, and frankly I'm glad to avoid another instance of cheap, stunted bananas and greasy motel "blueberry" muffins.

As we drive and eat breakfast sandwiches, Johnson tells me that he sometimes thinks of opening his own warehouse, probably in the Carolinas, to pack and repack whatever it is he's buying on his trips. But as we discuss it, he decides that a warehouse would turn into a processing operation, and he doesn't want to compete against U.S. scrapyards in the States. Right now, they're his customers. "What about you?" he suddenly asks. "Do you ever think of going into the business?"

Even the question makes me homesick. My grandmother passed a few months earlier, and for the first time I'm confronting the idea that even if I were to go into the business, there wouldn't be somebody there to bring in canisters of matzo ball soup for breakfast. I can't tell Johnson that, though, because then I'd have to explain my father. Instead, I tell him it's really up to Christine, my fiancée, and she doesn't know much about scrap yet. So we'll see.

South of Indianapolis he pulls the car into a brand-new multistory warehouse building just off the interstate. It belongs, I learn from the sign, to J. Solotken & Company, Inc. As we walk into the lobby, I feel like I'm in an insurance office, not a scrapyard. We're greeted by Brian Nachlis, an energetic, fortyish member of the family that owns the business, and he greets Johnson warmly. "We like Johnson and Homer," he tells me. "How long have we been doing business? Five, six years?"

Johnson relaxes in Brian's presence. There's none of yesterday's obsequious bended-knee tone. It's Johnson as I know him, confident and ready to deal.

Brian takes us out to the company's new three-story-tall, two-block-long warehouse. The perimeter is lined with neatly organized barrels, boxes, and bins of various kinds of metal, interspersed with machines and workers to process it.

This is a well-managed operation: as we walk, Brian leans over to pick up a Cheetos bag that he deposits into a waste bin. Then he pauses beside a barrel of scrap bathroom fixtures—faucets, mostly. They're brass, but they're also contaminated by pieces of steel and zinc. No U.S. brass maker will take them (and melt them) in such an unclean state, so they either need to go somewhere with labor cheap enough to break off and sort out those contaminant metals, or head to the landfill. "This is something I'd sell to Johnson," Brian tells me, and Johnson snaps a photo with his BlackBerry.

Next we stop beside several washing-machine-size boxes filled with water meters. It's the sort of scrap that's dropped off by utility companies, along with more valuable scrap like copper cables. But if J. Solotken wants the good scrap, then it needs to take the low-grade scrap, too. Water meters are classic low-grade: they contain some nice copper, but it's trapped in a case that needs to be broken to get at it. Johnson snaps a photo, and Brian picks up a water meter with two hands and passes it between them. "Before getting educated by Johnson, we used to throw those in the trash," he tells me. "But also remember," Brian adds with a smile at Johnson, "copper used to be sixty cents per pound."

"More than three dollars today." Johnson laughs. "So of course people pay more attention to the copper scrap!"

Later, in the car, I ask Johnson if it's true—that he and Homer really educated J. Solotken on water meters.

"Of course! Many U.S. scrapyards throw away good scrap that we can buy for China. It's part of the job to educate them in what's good. Then you become partners in handling the material." He's in a good mood: Brian sold a $50,000 container load to him, and he calls Homer with the news.

Meanwhile, the GPS directs Johnson to our next appointment. After that, it's lunch at a Chinese buffet Johnson likes on the north side of town, and then back down to Lexington, Kentucky, three hours away.

"We'll stay in Cincinnati tonight," Johnson says. "We have an appointment there in the morning."

I glance at him: this is his job for six months every year. It feels endless, and it is: there's always a scrapyard with more scrap to sell, always a factory in China that needs something to melt into something new. If Johnson doesn't bridge the gap, somebody else will. I glance at him and recall that he's not the only Chinese buyer taking a scrap-metal road trip today.

From Cincinnati, Johnson tells me, we'll drive a few hours to Canton, and then plunge south for hundreds of miles to the Carolinas. "I'd like to go to Cleveland, too, but no time. Anyway, it's better I go there with Homer. He knows that material."

CHAPTER 8

Homer

A month after my week-long road trip with Johnson, I travel to Guangzhou to visit some scrapyards for this book. I was hoping to meet up with Johnson's buyer, Homer, during this trip, and Johnson—claiming that I was his "good luck" charm during our travels—graciously arranged the meeting. As it happens, I'd met Homer briefly a couple of years earlier, but we were in a group and didn't have a chance to talk.

The meeting is set for late on a Saturday morning at my hotel. Wing Lai, Homer's son, waits for me in the lobby. He's in his early twenties, chubby and cheerful, and an important English-language-speaking member of his father's business. We shake hands as Homer emerges from the revolving door, hands in the pockets of his black windbreaker. Lai Huo Ming—that's Homer's real name—is a modest man of modest height, high cheekbones, full lips, and a perfectly layered, full head of hair. Twenty years ago, before making his fortune in scrap metal, he was a barber. Today, like most days, he still looks as if he just stepped away from the shop for the day. Homer doesn't excite people; he relaxes them. "You look a little fatter than last time," he tells me, his smile widening.

I shrug. It's the sort of thing that Chinese say without malicious intent.

Today, as always, Homer projects unflappability, and for a high-stakes commodity trader who only plays with his own cash, that's not a bad way to be. In that sense Homer is unique: most of the independent Chinese scrap traders I've met over the years are high-stress chain smokers who can't wait for dinner and some high-proof alcohol to cool the scrap-metal twitch. But in most other ways Homer is the quintessential Chinese scrap man, self-made and endlessly upwardly mobile.

"Shall we go?" Wing asks. We walk out to Homer's dusty black Honda with tinted windows. Most Chinese scrap men of Homer's worth (and even of much less worth) employ drivers, but it wasn't so many years ago that Homer was driving around Guangzhou on a bicycle, and then a motorcycle, in search of scrap metal to buy from rich Taiwanese importers. Like many self-made men, he's not comfortable with anybody else controlling his speed.

Wing settles into the passenger seat and slips an iPad from his shoulder bag. "This is our GPS," he laughs and pulls up a map. As our position appears on it, he turns to his father and—in Cantonese—directs him toward the highway.

I briefly met Homer in 2009, not long after the 2008 global financial crisis precipitated the quickest and most precipitous decline in scrap-metal markets in history—when American and Chinese consumers stopped buying new stuff, the price of raw materials crashed—cutting some scrap grades by as much as 90 percent in a matter of weeks. I wouldn't have known he was sweating, though: he was as calm then as he is now. Later I was told that he lost close to half of his considerable personal fortune in the crash. But all was not lost: within eighteen months, I've been assured, he had earned it all back—a reflection of his talents as a scrap trader, his charm as a salesman, and—above all—China's insatiable demand for metal.

Today, if market conditions are right, he and Johnson can easily buy and sell fifty or more shipping containers, worth of scrap metal per month. Some of those containers are worth as much as $100,000, and some bottom out around $10,000. They often take as many as six weeks to reach China from the United States, and Homer has the resources—and the courage—to keep buying even in choppy markets that can reduce the value of a container of scrap by 40 percent in the space of time it takes to travel from St. Louis to Qingyuan, Homer's hometown.

That's how good he is.

Guangzhou spreads out to our right. I can see the twisted Canton Tower, at nearly two thousand feet the tallest structure in China. As we drive, my eyes grow sleepy watching the long, monochromatic warehouses go by, each containing a factory, a product, a piece of China's economic engine.

"You see there?" Wing asks.

I turn to the thin concrete platform upon which high-speed trains will travel between Guangzhou and Wuhan, 630 miles north, in four hours. The pillars are still being poured, and in places the track, or even its concrete bed, hasn't been laid, yet. But it will be: since 2007, China has opened more than two thousand miles of high-speed rail lines. Each of those lines, it's not lost on Homer or Wing, requires steel for the tracks, and miles and miles of copper and aluminum for wiring.

After an hour Homer drifts onto an exit that leads into Qingyuan. At the top we face an under-construction housing development, dump trucks, and flatbeds carrying structural steel. Out the window I see farmers carrying crops on their backs, and the backs of little one-person pickup trucks. Homer gently passes them in the Honda, bringing into view a truck weighed down with shiny copper wire glowing in the sunlight. In the lane next to it is another small farm truck overloaded with snaky lengths of insulation, sliced open and stripped of its copper cable. As we pass, it flaps in the breeze, held down by rope.

Homer tells me that Taiwanese scrap men brought the recycling business to Qingyuan in the mid-1980s. Laborers were plentiful and willing to work for low wages that—all things considered—exceeded the subsistence-level living earned working in the fields. Better yet, Qingyuan was remote enough to avoid scrutiny from environmental authorities in Beijing and Guangzhou, yet still quite well connected—via rivers and railroads—to the ports where scrap was delivered. Those same links connected the scrap processors to the manufacturers who turn scrap metal into the wire, cable, and infrastructure that's driven China's growth for twenty-five years. Guangdong's government, eager for foreign investment and cheap raw materials, encouraged the trade. Pollution was an afterthought.

On the right there are run-down holiday hotels that advertise access to Qingyuan's famous hot springs and—just past them—a massive condo development behind signage announcing, in English, BADEN

SPA. Six-foot-tall photos of Europeans frolicking poolside run along the quarter-mile fence. Beyond it I see rolling hills and, in the distance, towering mountains. The hills are tree-covered, and then they're not— here and there somebody's dug into them, exposing the red soil to extract sand for construction projects. We pass shops wholesaling wire and cable for the region's booming housing sector, and as I gaze past them at the more distant hills, it occurs to me that this region was once painfully pretty.

When I ask Homer how and why he left hairdressing to get into scrap, he shrugs. "I just followed people into it." Family lent him roughly $5,000 to get started, and he traveled to nearby Dongguan and Shenzhen to buy his first loads from Taiwanese traders. Today, those two cities are the left and right ventricles of China's manufacturing heartland, driving growth throughout Guangdong Province and China. But back then they were just-emerging boomtowns where a smart and ambitious hairdresser with a tolerance for risk could get started on his fortune. "At the time, scrap is easier, cheap," Homer tells me. Back then, a truckload of old electric motors or refrigerator compressors imported from the United States could be had for around $1,200. It was during those early years that Homer learned, firsthand, the difference in copper content between, say, a three-quarter-inch cable covered in green insulation and one covered in black insulation and lined with steel. It's why, when Johnson sends photos from American scrapyards in the middle of the night, Homer need only glance at them once before typing out a price and going back to sleep. It's the kind of expertise earned, as much as learned, by stripping cable on your own.

Homer stops the Honda at the main gate outside Qingyuan Jintian, one of the largest copper scrap recyclers in China, and one of Homer's biggest customers.

A scrawny guard steps out, leans over the windshield, and, recognizing Homer, waves us through.

Qingyuan Jintian handles more than 90 million pounds of copper-bearing scrap per year—or roughly six shipping containers per day. Not all copper-bearing scrap is the same, of course. Some, like Christmas tree lights, contains 28 percent copper; some contains more. But the goal is always the same: take something that costs $0.55—like a pound of Christmas tree lights—and turn it into something that costs $3.12—the London price for a pound of pure copper at the precise hour of my visit. Qingyuan Jintian is governed by the same prices and principles that govern scrapyards across China and the developing world; the only difference is that it's much, much bigger. But already I know something for sure: whatever I'll see here is something that I've seen somewhere smaller.

Homer parks, and as we step out of the car a short, stocky young man with a baby face and an uneven homemade mop top (does Homer notice?) walks out the glass doors. He wears a black leather jacket that flares out behind him and the cocky sureness of somebody who was guaranteed his place in life. Which, in fact, he was: he's the nephew of the company's chairman. Behind him, and through the glass doors, I can see his secretary: a frail beauty in a miniskirt and black heels. I'd say she's out of place around here, but I've met enough guys like The Nephew to know that from his perspective, she's exactly where she belongs.

Homer and The Nephew chat warmly in Cantonese, and I look around the empty office campus. It feels desolate, as if there's nobody here. The only signs of life are a couple of uniformed men working intently on the engine of a truck parked beside a warehouse. Aside from Homer's voice, there's little to hear except, briefly, the grind of a distant machine.

A young man barely out of his twenties, wearing a gray uniform, pulls up in a golf cart. It has three rows of seats, and I take a seat in the last one, beside Wing and behind Homer and The Nephew. We're whisked around the driveway, past a truck arriving with an overseas shipping container and down a wide road. There's a warehouse on the left, and as we zip past I see the warm glow of shiny, shiny copper wires, pressed together into a bale. It's probably off to one of the company's furnaces elsewhere in Qingyuan, where it'll be melted and then sold to a manufacturer for melting into new products. On the right is another warehouse, this time darker, where people are clustered on their haunches amid piles of wire, and then they're gone, left behind.

Ahead there's a pile of hundreds of American traffic signs, green and white, nouns and verbs—TURN, PLACE, MERGE, HIGHWAY, KANSAS. I wish we could stop so that I could get a better look, just to know where they come from, but I suppose it's enough to know that they come from America, where alloys of aluminum and copper aren't needed quite so badly as they are in China.

Suddenly I'm hit by the scrapyard smell, that tangy metallic bouquet that brings me back to my grandmother. She would know exactly where we are without even having to open her eyes. And if the smell didn't tell her, the *clank-clank-clank* of a hammer against metal would.

We blast through a loading dock door into a capacious warehouse hundreds of feet long. On the left are large, dark piles of electric motors, and sitting beside them are six men on short plastic stools, wielding hammers to break them open. It's a scene repeated perhaps twenty times in this space—men and motors. We slow briefly, and I watch as they use pliers, pincers, and screwdrivers to pull out thin copper wires that pile up like red hair on a barber's floor.

Then, just like that, The Nephew's golf cart does a hairpin, speeds out of the warehouse, across an open space, and stops at the loading dock door to another warehouse. This time The Nephew suggests that we step out and take a walk. I'm happy to do so: sprawled before me, for hundreds of feet, are piles of loose cables and wires, some waist-high, some chin-high, in a sort of postapocalyptic high-tech moraine. Stationed in the valleys are teams of middle-aged men and women. The women run individual strands of cable and wire through table-size machines that cut a slice down the length of the insulation. As the strand emerges on the other end, the men grab it and pull away the insulation along the incision. The pure wire then goes in one pile, and the insulation in the other.

I've seen machines and wire-stripping operations like this on four continents, and in countries both rich and poor. A few weeks before I visited Homer, I visited one in St. Louis with Johnson; a few months before, I visited several in India. It's a simple process, neither high- nor low-tech. Rather, it's just tech, albeit the lowest-cost tech available to extract wire from insulation (so long as the wire is thick enough to be run through a stripper).

What's interesting is that, if you walk through this warehouse and ask a team member where her sister is, she'll point to the woman across from her; if you ask where her husband is, she'll point to the man pulling apart the wire she runs through the stripper. And if you ask where her parents are, she'll tell you the name of a small village in another town where the best economic opportunity remains a life stuffing seeds into the ground for little more than subsistence wages. At Qingyuan Jintian she can make around $400 per month—sometimes more—and if she pools that money with her relatives, they'll soon have enough cash to build a home back in the village and pay for a child's education.

Homer isn't thinking about that, though. Instead, he's looking at piles of bread-loaf-size cable that's been sliced open and stripped of its copper wires, only to reveal an equally rich prize: the bright copper foil that lines the insulation. It's beautiful, this shiny meat hidden inside the tough black husks. Someone just needs to mine it.

The Nephew tells us that he's awaiting a visit from a big German scrap company, and we need to speed up the tour a bit. I look away, distracted by a ten-foot-tall wall of printer cables, ribbon wire, computer mice, USB cables, and other information-age detritus compressed into La-Z-Boy-size bricks. It's like standing before a cliff filled with fossilized shells, except the fossils here are five years old and snaky. I spot a Microsoft mouse I once bought for $29.99 at a Best Buy; I see old printer cables that would've cost a ludicrous $39.99 each at the same store.

There was a time even five years ago when these cables would have been untangled, tested, and prepared for resale in used electronics markets around Asia. But Chinese consumers are richer now and, like their American counterparts, more and more prefer to buy new. What might have once been reused in China is now chopped to bits and remelted.

"Over here." Wing directs me around the cliff to another set of four conveyors that deliver wires too small to strip—things like mouse cables, USB cables, and thin ribbon cables—into upturned VW Beetle-size chopping machines. The choppers groan as the copper and insulation hits the cutting blades, and then—on the opposite end—those cables emerge as tiny bits of rubber confetti and copper. From there, they drop onto large shaking water tables—the same kinds of tables that Raymond Li uses to recycle imported U.S. Christmas tree lights. The water washes the plastic and rubber in one direction; the heavier copper slowly moves in the opposite direction, like heavy stones in a flowing river. These water tables wouldn't work on Raymond's Christmas lights, however; instead, they're perfectly tuned to separate wire from plastic in the cables that Americans toss out when they get rid of their old PCs.

The results, however, are much the same as Raymond's: clean copper fragments, ready for market. Wing nudges me and points at giant sacks of chopped copper—4,000-pound sacks, just like the giant sacks of copper I saw at OmniSource in Fort Wayne, Indiana—ready for sale to somebody who makes things from copper. Of course, this isn't nearly as sophisticated or as large as the Brontosaurus of a chopping line that OmniSource runs, but it operates on the same principle: reduce the wire to its smallest bits and then use available technology to separate, separate, separate. It's just that Qingyuan Jintian's technology of choice is one utilized in hundreds of scrapyards across China: a shaking water table.

This raises a question: Qingyuan Jintian is surely wealthy enough to purchase the technologies available to OmniSource, so why doesn't it? The simple answer is that they might, soon: in 2011 the Chinese government financed the building of a high-tech chopping line in Zhejiang Province. The more complicated answer is that there's still a lot of human labor involved in running Qingyuan Jintian's line. Even though the company uses mechanical means to chop up and separate cables, it still pays employees to snip a mouse from a mouse cable (there's a separate market for the dead mouse), a steel plug from a printer cable (there's a separate market for that, too), or a steel USB plug from its USB cable (it, too, has a separate market—around 2 cents per pound). Each of those snips—the mouse, the USB port—means that there's less possible contamination (steel, in particular) in the copper that the company markets to its customers. Without the human element, the company would need to employ magnets, and those can't be guaranteed to do the job as well as a human being with scissors. Hand labor means that objects are recycled more completely—and more profitably—than they would be in the high-tech developed world. Sustainability, in this case, parallels profitability.

"Where does the plastic go?" I ask Wing.

Wing leans over to The Nephew and yells my question into his ear. The Nephew raises his brow and nods at an open loading door and a long rectangular pool, roughly half the length of the warehouse. Inside is the water that the company recirculates to the water tables, and tons of rubber and plastic insulation.

"They hold it here until it's time to ship to a company that wants it," Wing says.

"Is it worth anything?"

"Around two hundred to three hundred yuan per ton." That was $31 to $47 per ton on the day that I visited, or roughly the price of electric motor scrap back in the late 1980s.

"Are you hungry?" Wing asks. "It's a good time for lunch, I think."

I glance at Homer and The Nephew. They're busy talking in Cantonese, so I take a moment to walk back into the building and count the people. There might be twenty. Most are women—it's hard to tell with the baggy clothes and face masks—and they're moving methodically, slowly chipping away at the cliff of printer, USB, and ribbon cables, reducing all those overpriced computer accessories into peppercorn-size flecks that can be made into new overpriced accessories. As China becomes wealthier, more and more of that scrap is not going back to the United States but rather staying in China as new products, for sale in the ever-expanding shopping malls of Shanghai and other wealthy municipalities.

"What do you think?" Wing asks me.

"I think China has really figured out how to make money off the United States."

"Yeah?" He laughs.

"Yeah. They call this dumping in the States."

"Dumping? I don't know the meaning."

I want to tell him it means "dumping your waste on poor people so you don't have to pay for its disposal." But I'm not sure that would make much sense at the moment. "Never mind."

Homer Lai grew up in a small farming village, but today he lives near the top of a luxury high-rise overlooking the Bei River in Qingyuan. It's a big place—four bedrooms—but the decor is functional and understated. The walls are mostly bare; the furniture is big, comfortable, but by no means expensive. The most prominent feature might be the large flat-screen television. After that, what I really notice about the place is the family: Homer's elderly but utterly vital mother has her own room; his son, Wing, and his pregnant wife occupy a room at the opposite end of the apartment; Homer's strongly built wife wanders out of yet another room. Downstairs, in another unit, is his sister. She comes in and out of Homer's unit as if it's hers. As Johnson told me, Homer likes to keep his family close.

Then there are the windows.

The view from the front of the apartment reveals the river, riverbank high-rises, and a growing city that extends out into the villages and towns where scrap is processed into copper. But it's the view from the back of the apartment that surprises me. A city that sprawls like Los Angeles extends out to the distant mountains, where even now businesses recycle the worst of what they import in hollows out of sight and mind of regulators—and their city neighbors. Much of that burning is devoted to high-tech scrap, such as silver-coated wire, a key component of many high-tech devices. Years ago the United States had refineries that could separate the copper from the silver, but environmental problems shut them down. Now the silver-coated wire ends up in the hills, where it's processed using acids away from the eyes of regulators. By day, the silver and the copper comes down from the mountains, to be incorporated into new products shipped all over the world.

But none of that's obvious from Homer's balcony. Spread out before me, most of Qingyuan's buildings are no more than ten stories high, and they roll out along boulevards that intersect at clusters of thirty-story high-rises. I'd always thought of Qingyuan as a small town, verging on a city; what I see is a metropolis. Who knew? As I gape, Homer reminds me that most of these buildings were wired with metal imported as scrap, and processed locally.

Thank you, America, I think.

I'm offered a seat in one of two large leather chairs at opposite ends of a marble coffee table. Homer sits on the brown leather love seat next to me and sets out several small cups for tea that he boils in front of us. Wing joins in the chair opposite of mine, and his pretty, very pregnant wife takes a seat on the arm. Homer picks up a remote control and starts a DVD. "My son's wedding," he says. The first sets of images show a caravan of thirty-five cars that carry Wing to pick up his bride. They stop, and dozens of people emerge, including—I see right away—Johnson.

I ask Homer if he had thirty-five cars in his wedding. "I had a motorcycle," he tells me with a laugh. "We went to the marriage bureau on that."

Wing's wedding lasted for three days, with multiple banquets and hundreds of guests. Homer watches it with quiet contentment; Wing beams.

"Are you hungry?" Wing asks me.

I am. But first there's something that I want to see. "Homer, can I see the computer where you look at Johnson's photos in the middle of the night?"

He turns to Wing and smiles. "He wants to see that?"

Maybe I shouldn't have asked.

"Come," he says, standing and beckoning me to the back of the apartment.

We walk through a doorway, and I'm surprised, and a little embarrassed, to find myself in a bedroom—Homer's bedroom. If I had known, I wouldn't have asked. I just imagined that he was in a study somewhere. It's a plain, totally unremarkable space that wouldn't have been out of place in a small village home—which perhaps is the point. The full-size bed has a simple wooden headboard and is covered in a rattan mat. Otherwise, there's just a stand-up fan, a twin fold-out bed, and a built-in desk next to a balcony door. Homer's Lenovo laptop sits atop the desk, next to a vanity mirror that I presume belongs to his wife. A lemon sits between them.

"This is where you work?" I ask.

He opens the laptop and clicks open his incoming e-mail folder. As I watch over his shoulder, he pulls up photos that I watched Johnson take with his BlackBerry. I'm suddenly reacquainted with the Christmas tree lights from Cash's Scrap Metal & Iron and the water meters from J. Solotken. He clicks through several dozen additional images until he slows and reaches a tangle of wire that we saw in South Carolina. Then he slips on his glasses, peers at it carefully and taps the screen. "The green wire, that's maybe sixty percent recovery. The red is lower, maybe forty percent."

I look at him, and then look out the window. Out there, somewhere, is a sun-baked patch of ground where he learned, using his own hands, how much copper is to be found in the wires of different colors, and how much it costs to extract it. Compared to that, and even barbering, this is a fine, fine life. If I could live it, I wouldn't hesitate. "You don't mind getting up in the middle of the night to look at these when Johnson sends them?"

"Why not? It doesn't wake my wife, and then I just go back to bed."

His family crowds around the computer, discussing the images he's flipping through. Everybody knows a little something about the family business, everybody can comment on whether a load of wire is worth the price. I step back and take a photo: a scrap man's family portrait.

Later, looking at the image, I notice newly built high-rises through the sliding door beside the desk. One day not so long from now, those high-rises will consume goods with the gusto of their New York—and Shanghai—counterparts. Whether they will waste as much is a question that nobody can answer yet. But I have no doubt that Homer, and especially the children crowded around him, will soon worry more about the scrap generated in those buildings than about the photos of scrap taken in the middle of the night on the other side of the Pacific Ocean.

CHAPTER 9

Plastic Land

A city of twenty million people generates a lot of trash. Some of it ends up in landfills, and some of it ends up being recycled. Beijing, a developing city of at least 20 million, recycles more than most, in large part because it's home to millions of migrant laborers, many tens of thousands of whom make a living by buying and sorting the value from what their upwardly mobile neighbors throw away.

The migrant peddlers aren't hard to miss. They ride tricycles retrofitted with trailers filled with what most Beijingers consider junk: newspapers, plastic bottles, bits of wire, boxes, and old appliances like televisions. Sometimes they stop at garbage cans to dig for what might have been thrown away; more often they take house calls from building guards who notify them of a resident up in a high-rise with a big cardboard box that held a new HD television, and some beer bottles to sell.

Over the years, a handful of Chinese academics have attempted to quantify how much trash and recycling Beijing generates on an annual basis; they've failed, roundly. The business is so big and yet so lacking in organization (it's largely conducted by migrants who don't pay taxes, and prefer to remain anonymous) that it's impossible to add up. Nonetheless, it is possible to figure out where most of it goes.

Enter my friend Josh Goldstein, a professor of contemporary Chinese history at the University of Southern California.

Ten years ago, while sitting in a Beijing library, boning up on Peking opera, he noticed scrap peddlers walking past his window carrying all manner of waste and recyclables. "So one afternoon I just decided to get up and follow them," he told me. "And I ended up at this massive recycling market. I started working on the subject from there." Along the way, he traced out the history of how Beijing recycles, and managed to locate the factory responsible for recycling all of the plastic cups that KFC generates in the city.

Josh is smart, sharp-tongued, and adventurous. In mid-June 2010, one of his connections offered him the chance to see what was being described as "the place where Beijing's plastics go." He agreed right away and called me up shortly thereafter. "Wanna come? Not sure what we'll see, but it's worth a shot, I think. I have some people who can take us around."

The place is called Wen'an County.

I didn't hesitate.

In Beijing we catch an early-morning minibus south out of the city via two-lane roads that skirt the tollways. Two hours later we're dumped at a rural gas station wedged into a dusty crossroads. The crossfire truck traffic that travels through it is deafening, and the exhaust it kicks up is stifling. Some of the trucks pull empty trailers, some carry drywall for construction projects. But most are loaded with scrap plastics: auto bumpers, plastic cartons, and giant ugly bales of mixed plastics ranging from shopping bags to detergent bottles, Folger's coffee cans to food wrappers. Few American recycling companies will accept this last category—at least, they wouldn't in 2010—but many American recyclers place them into their recycling bins anyway, and some recyclers, who'd rather sell them than pay to landfill them, offer them to scrap brokers with customers in China.

Still, all of that household recycling comes as a bit of a surprise: Josh had mentioned to me that Wen'an County imports plastics from abroad, as well as Beijing, but I wasn't expecting to see what basically amounts to my mom's trash riding through town. But in retrospect, that was just my scrap-metal-centric shortsightedness getting the better of me. If my travels in global recycling have taught me anything, it's that somebody in the developing world can usually find a use for what Americans can't recycle profitably.

According to the China Plastics Processing Association, in 2006 China was home to roughly 60,000 small-scale family-owned workshops devoted to recycling plastic, according to the most recent statistics, good or bad, that my government sources will give me. Of those, 20,000 are concentrated here, in Wen'an County. In other words: Wen'an County isn't just the heart of northern China's scrap-plastics industry; it *is* the Chinese scrap-plastics industry. And because China is the world's largest scrap-plastics importer and processor, I think it's fair to say that Wen'an County is at the heart of the global scrap-plastics trade.

I glance at Josh: he's lanky, with a full black beard and a backpack that makes him look like he's arrived here via *Lonely Planet*. He's traveled widely in China, speaks the language, and knows what he likes. And he doesn't like this gas station. Fortunately, our connecting minibus arrives, and we're on our way.

Before long, the single-lane road is overcome with dust and garbage. Traffic is choking with trucks weighed down by refrigerator-size bales of imported old plastic; on both sides of the road single-story one-room workshops are bathed in a swirl of dust. I note that the shops are covered in brightly colored signs advertising the two- and three-letter abbreviations for the various grades of plastics bought, sold and processed in the county: PP, PE, ABS, PVC. Polypropylene. Polyethylene. Acrylonitrile butadiene styrene. Polyvinyl chloride. Those letters all look so exotic, so distant and industrial. But they're not: they are the formulas that make up the plastics that encase my phone, my coffee, my laundry detergent bottle. This is the stuff that my friends and family dump in their recycling bins.

As I watch out the window, it looks to me as if nobody in Wen'an landscapes their storefronts. A few businesses might leave out piles of old taillights and bumpers, in cases where there's no room in their warehouses, but most use their storefronts to dry piles of wet, shredded plastic. It's a bustling, crowded, and incomprehensibly dirty main street, crossed by the occasional stray dog, partly blocked by a broken-down truck, and frequently scarred by black spots where—I'm later told—unrecyclable plastics were burned in the night. Above me, plastic bags are captured by the wind, floating on the breeze. But what I find most striking about Wen'an is this: there's nothing green. It's a dead zone.

As we drive, I look through an open door and see shirtless workers feeding red automobile taillights into machines that chop them up into fingernail-size flakes. Through other doorways, I see the air shimmer with hot fumes. "This is really it," Josh sighs as he gazes out the window. "What a fucking shit hole."

Our first stop is our hotel, and suites as large as parking lots, outfitted with beds the size of tractors, and covered in carpet the depth of an American lawn. I don't know downtown Wen'an, but I know enough about China to recognize this as the kind of place where public officials go when they can only get away from their wives for a few hours. For all of the dirt and garbage out front, this is a good reminder that somebody, somewhere, is making money in these parts. And yet, out my window, a woman is busy picking plastic bags from trash in a small brick court-yard. Beyond her is a series of red-roofed warehouses that gives way, eventually, to an under-construction twenty-story high-rise, rising like a single candle on a rotten birthday cake.

It wasn't always this way.

Twenty-five years ago, Wen'an was bucolic—an agricultural region renowned for its streams, peach trees, and simple, rolling landscape. The people who knew it then sigh when they recall the fragrant soil, the fish-ing, and the soft summer nights. Engage a local in conversation, and within minutes you'll hear how you should've come in the old days, back before the business of Wen'an was the business of recycling automobile bumpers, plastic bags, and bleach containers, back when the frogs and crickets were so loud they drowned out human conversation, back be-fore the development of the plastics recycling trade plasticized the lungs of men in their twenties, way before multinational companies did busi-ness in Wen'an so they could say their products were "made from recycled plastics."

Then China started to develop, and a brisk and growing demand de-veloped for the plastics that go into new buildings, cars, refrigerators, and all the stuff people buy. Most of those plastics were virgin, made from oil. But that was a temporary state of affairs: the stuff people bought became the stuff they threw away, and soon there were enough scrap plastics in China to justify going into business to recycle them—and thereby compete against the virgin plastics manufacturers.

As recently as fifteen years ago, Wen'an's waste plastics industry was devoted almost entirely to recycling plastics generated in China. But demand for plastics was growing rapidly both in China, and outside of it, and by 2000 China's plastics traders were looking for additional sources of scrap plastics. They found those plastics abroad.

Then, as now, few American, European, or Japanese scrap-plastics exporters have any idea who recycles the material that they export. Instead, they sell to brokers and other middlemen who sell to Chinese importers, often near ports, who then resell the scrap plastics to small traders of the sort who transport the plastics to Wen'an. Once the plastics arrive in Wen'an, they're sold again. By the time a bundle of U.S. plastic detergent bottles is bought by the family that will actually separate and recycle it, it's all but impossible to trace it back to the American families who might have thrown those wrappers, bags, and bottles away.

It's a shadowy trade: unlike the multibillion-dollar trade in recyclable metals, plastics are traded in small lots. Indeed, for all of the commerce that happens in Wen'an, the 450 square miles of the county and its 450,000 residents (as of 2004) remain nearly unknown outside the immediate region and industry. The local government—and arguably, the environmental authorities in Beijing—surely prefers it that way.

But for all of the uncertainty, one thing is absolutely certain: foreigners aren't welcome unless they're here on business. And Josh and I definitely aren't in Wen'an on business. If not for a well-placed connection (of whom the less written, the better), we wouldn't be here at all.

It's late morning when Josh and I reconvene in one of the hotel restaurant's private dining rooms. We're joined by a local who will serve as our driver, and a representative of a local recycling company. Josh makes small talk, never once mentioning that I'm a journalist. If they knew, I'm not sure we'd be welcome to stay.

Our waitress—her name tag identifies her as #200—is a demure figure in a red skirt and a matching coat two sizes too big. For lack of other interviewees (for the moment), we ask whether she knows anything about the local plastics industry. "PP, PE, ABS," is her immediate reply, as if listing the day's lunch specials. "My family does the business." Intrigued, Josh asks just what percentage of the county is actually involved in the trade. "Find out the number of households," she replies. "And that will tell you the number of businesses. And if you don't have enough money to start your own business you go work for someone else's."

In Wen'an, we learn over lunch, it's possible to enter the recycling business for as little as $300—enough to buy a used shredder to chop up anything from taillights to plastic WD-40 containers, a tub to fill with caustic detergent to clean the chopped plastic, and a truckload of plastics to recycle. Environmental and safety equipment is neither required nor available at the local equipment and chemical dealers (we checked).

Our driver, hunched over a plate of unpeeled shrimp, looks up. "I used to be in the business, too. Now my daughter is married to a guy in the business. ABS, PP, PVC."

The waitress nods. "I have two brothers in the business. The money is much better than what I can earn waitressing."

Josh furrows his brow. "So why, then, aren't you in it?"

"It's unstable," she explains with a shrug. "And the health effects are bad. It's not like it used to be around here." Like others we'll encounter, she recounts the paradise that was, as others have recounted it to her— the peaches so sweet, they could sell as candy.

The precise details of how Wen'an was transformed into a global plastics recycling center are lost to history. Still, as we chat with the locals, it becomes clear that it was accidental rather than a grand plan. "Someone started doing it," explains one knowledgeable local who has worked in the industry for years. "He made money, so more people did it. The government saw it as a good source of tax revenue, and encouraged the industry. It was random."

Another successful business owner tells us that he started out in the mid-1980s, buying plastic bottle caps that nobody wanted and holding on to them until he could figure out a way to process them into reusable plastics (I possess a vivid, totally invented image of his wife glaring at garbage bags full of useless bottle caps, bought with the family's life savings). Eventually he did, and by 1988 he and other entrepreneurs in the area were opening small processing plants. With money on their minds, and increasingly flowing into their pockets, the town's leaders turned their backs on the obvious negative effects of becoming a place where other people dump their trash—even if that trash has value.

In fact, Wen'an was the perfect location for the scrap-plastics trade: it was close, but not too close, to Beijing and Tianjin, two massive metropolises with lots of consumers and lots of factories in need of cheap raw materials. Even better, its traditional industry—farming—was disappearing as the region's once-plentiful streams and wells were run dry by the region's rampant, unregulated oil industry. So land was plentiful, and so were laborers desperate for a wage to replace the money lost when their fields died. As I hear these stories, I can't help but wonder: How much of the plastic that Wen'an recycles was made from the oil pumped from Wen'an's soil? Are all those old plastic bags blowing down Wen'an's streets ghosts of the fuel that used to run beneath them?

After lunch Josh and I are driven out of central Wen'an to visit a plastics recycling factory with two representatives of one of the county's biggest processors. Downtown's dust, grime, and billowing trash give way to gentle, rolling fields and the fruit tree groves for which Wen'an was celebrated. But only briefly: on the left is a fenced-in brown dirt yard piled with table-size bales of plastic scrap. Bags twirl and corkscrew into the hot summer breeze, catch flight, and tumble over fields until snagged in stiff, dead grasses. Behind the fence, two workers squat over a broken bale of plastic automobile bumpers, picking trash from its tight recesses, while another runs the bumpers through a shredder. The other garbage stuck to that bumper and bale is plastic, too, and it too will be segregated and recycled. In the United States, no recycling company could afford to pay somebody to do that, because the value of the plastics is too low. But even if it were profitable, there's another problem: the plastics that can be made from recycling wrappers and other cheap plastics don't meet the quality standards of U.S., European, or Japanese manufacturers. Only the Chinese, often manufacturers of last resort, will use that stuff.

As we bounce down the road, one of the company representatives tells us that most of Wen'an's plastics businesses are located in forty to fifty villages that spill across the rural, unconnected county. The small scrapyard behind us belongs to a village, one of the company men tells us; it's rumored to manufacture plastic bags from an ugly mix, including industrial-use plastics, which are then passed off as safe for food packaging. As the company men laugh at this, Josh looks at me—and then joins in, ruefully.

We cross a bridge over an algae-choked river into a tighter, dustier version of the streets of downtown Wen'an. Unlike those streets, however, these paths are crowded with groups of half-naked, often barefoot children who race and play among trucks piled with sheets of corrugated plastic boxes, old plastic barrels, and giant dried puddles of plastic that dripped onto factory floors, to be shoveled into containers that were exported to Wen'an. They look like fossilized heaps of cow shit.

There aren't any markets, restaurants, or even equipment dealers in this village. It's all makeshift warehouses, fence posts covered in tree bark, and open spaces piled with bales of bumpers, piles of plastic barrels, and stacks of plastic crates. Our driver turns at a corner warehouse covered in graffiti phone numbers and stops at a small office building, next to a shiny black BMW. Despite all of the industry we saw on our way here, it's quiet in the village, almost as silent as wilderness. A distant mechanical buzz is no more obtrusive than a birdsong.

As we leave the car, we're greeted by a man I'll call Mr. Hu, a fiftyish owner of the plastics business we are here to visit. He wears a large Rolex with his gray work jumpsuit; in the factory, I notice his employees wear shorts and, occasionally, shirts. He's handsome and well-fed; they're scrawny and bug-eyed. Across the dirt street, employees have just started a small plastics shredder that they use to reduce plastic fruit baskets that Mr. Hu imported from Thailand, into flakes for recycling.

Mr. Hu tells us that he's been in the waste business for twenty years, but that this factory is only seven years old. He owns 90 percent of it, and "investors"—often a euphemism for the local government—own the other 10 percent. He leads us into an open courtyard where five employees—three of them shirtless teenage boys—pick trash from a load of unidentifiable, partly shredded bits of plastic imported from the United States. I ask what it was before shredding, and Mr. Hu shrugs. "Maybe boxes. Maybe something from cars."

As we watch, the shredded plastic is poured into metal tubs full of caustic cleaning fluid, and washed by turning metal strainers through the mix. Then it's spread out on tarps to dry. When the workers are done, the excess trash and cleaning fluid is gathered up, and either resold or tossed into a waste pit on the edge of town. Unless I'm missing something, or visiting on the wrong day, there's no safety equipment, no respirators, hard hats, or steel-toed boots, here; in fact, most of the workers—including Mr. Hu—wear sandals.

I look at Josh, and he looks back at me: this is bad.

"We only have one extruder running today," Mr. Hu tells us. "In here."

We walk into a brighter room, roughly forty feet long and perhaps half as wide. It smells of something modern and chemical. In the middle is a long device that runs perhaps half the room's length. At one end is a worker who pours boxes of shredded plastic flakes into a table-size funnel, where they're slowly melted. I can see the heat—and the melted plastic fumes—rising into his face. Meanwhile, the plastics drip into a ten-foot-long pipe, eventually emerging as fifteen pencil-thin gray noodles. The principle isn't much different from the one used in a pasta maker. The only difference is that the plastic noodles are cut into quarter-inch pellets and packed into bags for sale to manufacturers.

At Mr. Hu's factory, conditions are actually better than most factories in Wen'an—or so Mr. Hu claims. Yes, sure, a worker stands above the machine, inhaling the visible fumes that fill the room with a chemical choke. But according to Mr. Hu, the company has taken some tangible steps to improve the situation. "We used to pay the guy on the extruder more. But that was before we improved the ventilation." He nodded at the open bay doors and the open windows above the shop. Now he earns the same as the poor wretches who wash shredded plastics in chemicals without the benefit of gloves.

Mr. Hu invites us back to his office and offers a seat at a large wooden worktable. Behind us, his wife is working, and his son is playing computer games on a PC. As he pours tea, Mr. Hu tells us that among the customers for his recycled plastics are two companies on the *Fortune* Global 500 list—one of whom is also on *Fortune*'s World's Most Admired Companies list. The other company awarded Mr. Hu's company a pass on its RoHS (Restriction of Hazardous Substances) evaluation, an industrial standard meant to require health, safety, and environmental compliance of contractors. To prove it, Mr. Hu produces the paperwork. As it happens, the phone in my pocket was made by one of the manufacturers named in that paperwork. I hold it up and ask: "Maybe the plastic came from here?"

"Maybe! Maybe so!"

Mr. Hu, too, has memories of Wen'an before the plastics. He grew up in Beijing, but because his mother was from Wen'an he often visited her family as a child. "I loved coming here," he says. "The earth itself was so fragrant. You could drink directly from the streams, and there were plenty of fish." He shakes his head with a sad smile.

"You can't put Humpty-Dumpty back together again," Josh whispers to me.

"What about the health effects of plastics recycling?" I ask.

Mr. Hu shakes his head. "You can't say precisely what the health effects are. But if you take a kid from a healthy environment and one from an environment filled with trash, the latter will be the one with problems." As I glance at his son, Mr. Hu adds that high blood pressure and other "blood diseases" are common in the area. But the biggest problem is the stress related to living in a "dirty, stinky, noisy environment. It takes a physical and mental toll." He reaches for my digital camera and holds it in his palm. "Is there a place to process and treat this when you're done with it? There's a law, sure. But if you ask somebody where to do it—no."

I'm not sure what, exactly, he's getting at. Perhaps he's suggesting that the pollution that's obvious all around us isn't really his fault. No doubt, a lack of government regulation played a role in the uncontrolled, unsafe expansion of Wen'an's recycling industry. But the decision to pollute, to ignore the safety of workers, ultimately rests with people like Mr. Hu. I glance at the Rolex on his wrist and the PC on which his son is playing video games. For the price of either, he could buy respirators to keep his employees safe from the plastic fumes that they're now inhaling. If he traded in his BMW for a Buick, the difference could fund work boots and jumpsuits, like his, to protect the entire village labor force from sharp edges, burns, and falling objects.

Josh purses his lips. "Do any of the business owners ever get into trouble for their activities in Wen'an?"

Mr. Hu shakes his head and explains that if a trader misrepresented his goods, he might get into trouble. But in his memory, the only health or safety violation that triggers government interest is when low-grade plastics are incorrectly marketed as safe for use with food. "Otherwise this [industry] is a good source of tax revenue. That's how they see it."

Wen'an has become modestly prosperous over the last two decades—at least at the upper end of the income scale. BMWs and Land Rovers are common on Wen'an's roads. But best as Josh and I can tell, that money has done little to improve the lives of the people who work in factories like Mr. Hu's. Wen'an's schools are so poor that families who can afford better—like Mr. Hu's—send their children out of the region at the first opportunity. Nobody wants to live in Wen'an—not even Mr. Hu. He has a home in Beijing, and that's where his wife lives most of the time, along with their son.

As the conversation begins to fade, Mr. Hu offers something unexpected: he asks if we'd like to see where he and the other village plastics business dump their waste.

Perhaps it's not nearly as bad as everything else around here? I can't imagine why else he'd offer, and Josh and I readily agree.

We leave town with the two company men in an SUV, driving down a muddy road interrupted by deep holes that look as if they'd been left in bombing runs. A quarter mile requires ten minutes; the landscape is dry and bleak. Then ahead of us I see rows of waist-high burial mounds. There are hundreds of them, dimples on the Chinese landscape, the final resting places of people who used to farm this region. It occurs to me that we're driving through a cemetery, not farm fields.

We turn right onto hard, cleared land that spreads out flat beside the mounds. Ahead of us are colorful streaks of plastic in the black dirt, traces of the plastics dumped into the giant pit spread out before me: it's at least two hundred yards long, a hundred yards across, and twenty feet deep. The dirt walls are streaked with trash, and its floor is filled with green and brown water swirled with colorful plastic bags. It is where, we are told, much of the village's plastic cleaning fluid and unusable waste is dumped when nothing else can be done with it. That goes for Mr. Hu's factory, too.

I look to my right, at the burial mounds, and notice that one of them has been severed in half and is slowly crumbling—bones and all—into the pit. The excavator that dug this pit cut cleanly through that grave as if it meant nothing to anyone, as if it were just dirt. It's shocking: in China, where reverence for the dead is among the deepest of cultural imperatives, that pit, literally etched into a cemetery, is a cultural transgression of the first order.

As we stare out at the canyon of trash and chemicals, Mr. Hu's employees go quiet. I don't know what they're thinking, but they don't look happy.

Josh glances in their direction and, with forced enthusiasm, repeats the Chinese Communist Party's canned response to whomever questions their commitment to the environment: "There's been a lot of economic progress out here," he says in Chinese. "Price of progress."

"Yeah," one mumbles, kicking at dirt that very well might have covered his ancestors.

Every morning, seven days per week, hours before dawn, trucks arrive on a wide, quarter-mile long side street that peels off downtown Wen'an's main street. It's a striking transition, even from the cluttered, dead grime that is Wen'an's main boulevard. Out to the horizon, tractor trailers and pickups are piled with a wild assortment of scrap plastics for sale to the small, temporary tables and stalls that sprout on the broken, burn-scarred cobblestone pavement. According to numerous market participants, roughly 70 percent of the scrap that arrives here every day is imported, driven from the ports overnight. The remaining material—usually the poorest quality compared to the nice stuff the less thrifty Americans and Europeans toss out—comes from nearby cities like Beijing.

The market is a wild place: pool tables are set up beside bags of brightly colored plastic pellets; traders play cards for loads of plastics that cost thousands of dollars. Children play between the trucks, the piles of plastics, the heaps of garbage, and the men seated on canvas bags of recently recycled plastics ready for sale to whomever is making something from plastic today. The action peaks around five in the morning as the overnight trucks arrive to unload, and dwindles to the lowest-grade materials, and the smaller dealers, by seven.

We arrive late—around six—but the upper section of the street is still dominated by a long flatbed piled ten feet high with tightly packed automobile bumpers, laundry detergent bottles, plastic washing machine gears, plumbing, defective factory parts, television cases, and heavy-duty plastic bags stuffed with plastic factory rejects from somewhere far away. Workers climb atop it and unload the pieces by hand, dropping parts and bags to the ground, where they're inspected and weighed by two portly men with notepads. As we watch, our driver tells Josh that there were 120 metric tons of plastic on the trailer (a wild exaggeration), and that he makes the trip three times per month from Harbin, a city roughly six hundred miles away.

We walk the length of the street, through dozens of salesmen, past a county-run scale that—the operator tells us—weighs a hundred loads of plastics per day. The cobblestones bake in the sun, covered in trash, melted plastics, and burn marks where unrecyclable—that is, unsalable—

materials were dispatched in the night. Here and there, small-scale buyers cart around old plastic detergent containers dripping of their former contents; the pungent aroma of melting plastic wafts through an open gate. At the end of the street is a drainage ditch—perhaps once a creek—choked with garbage, a plastic mannequin head, and the remains of a green plastic bin with three circling arrows and the word RECYCLING in English.

Wen'an is the most polluted place I've ever visited. I can't quantify it with data, because nobody has ever taken the data. But the scale of the pollution, covering much of the county's 450 square miles, is unmatched anywhere else I've been, in any other country on earth.

So what's to be done? Can't somebody in nearby Beijing shut it down?

Almost two years to the day after Josh and I visited Wen'an, I received an e-mail from him with some surprising news: Wen'an's new Communist Party secretary had ordered the total shutdown of the county's plastics recycling industry. Later press accounts claimed that 100,000 people were immediately left jobless, and countless thousands of small family businesses were rendered effectively bankrupt (both figures are believable). My initial response was a rush of jubilation: if anything needed to be shut down, it was Wen'an.

I really should have known better.

I flew up to Beijing a few weeks later and learned that—in the wake of the shutdown—the price of scrap plastics in the city had fallen by half. Peddlers who spent their days collecting plastic scrap suddenly had much less of a reason to do it. Warehouses where plastics used to be sold before being transferred to Wen'an started overflowing; plastics that used to be pulled from the trash, and people's homes, now stayed in the trash.

But the more serious problem was the long-term one. China needs recycled plastics to make everything from cell phones to coffee cups, and shutting down Wen'an won't make that demand go away, any more than plugging an oil well will make people stop wanting gasoline. Wen'an's plastics merchants know that as well as anyone, and in the wake of the shutdown they scattered over northern China, reestablishing their unsafe, unclean operations wherever they could find an amenable government. What had once been a disaster spread across one county is now a disaster spread across northern China.

Who is to blame?

For sure, blame in part belongs to China's regulators. Despite the

Chinese government's popular image outside China as an all-powerful, uniform, centrally administered force, it actually has very little influence at the local government level. Nonetheless, even if it were as organized and powerful as many in the developed world think, it's not so powerful that it can instantly transform the world's largest recycled plastics industry into the world's cleanest recycled plastics industry. Doing so would require figuring out something that Europe, Japan, and the United States have all failed to solve: how to recycle, profitably, all that ugly mixed plastics.

But as Josh pointed out to me in an e-mail, fixing Wen'an didn't have to entail wholesale industrial change. "No serious effort was ever made to work with the thousands of small-scale mom-and-pop processors [in Wen'an] to solve [environmental and safety] problems." Simple steps— work boots, respirators, and a municipal wastewater treatment system— would have made a big difference.

Ultimately, I believe, blame must also be placed on the consumers and the home recyclers—in China, in the United States—who buy plastics and then dispose of them in ever-greater quantities in their recycling bins. To be sure, it's all but impossible for a home recycler to determine where, precisely, a bin of recycling ends up. But it *is* possible to prevent that bin from filling in the first place. Don't like Wen'an? Worry less about where your garbage man is taking your trash, and more about how much of his truck it fills. In the meantime, the corporations that support Wen'an and places like it by purchasing recycled plastics from its businesses would be well advised to seek out cleaner raw materials. Sooner rather than later, enterprising journalists are going to figure out how to connect their demand with their polluting suppliers.

Perhaps at some point a private company will figure out a way to recycle all that cheap plastic. Or perhaps a government will do it first— China is spending millions on recycling research; the U.S. government is funding very little. It wouldn't be the first time that technology and the scrap industry have combined to bail consumers out of their waste. But until that theoretical bailout comes, the world may have to learn to accept the reality of Wen'an and its progeny.

Before leaving Wen'an, Josh wants to speak to a doctor or some other medical professional about the health of Wen'an's people. So, late in the afternoon we wander into some of Wen'an's few remaining village-style lanes, in search of a clinic. Our odds aren't so bad, actually: most small Chinese villages have a nurse or village doctor who can handle the minor medical emergencies of daily life.

Soon we come across a colorful tiled gate, and just beyond it, a pleasant courtyard. We stroll inside and, spotting an open door at the far end of the space, wander into a small office where a pudgy, solidly built late-middle-aged man is seated at his desk in flannel shorts, a gray polo shirt, and sandals with black socks. Light comes through the door and from a desk lamp, but it's a shadowy place. Two beds are set against the back wall. They're covered in old, dirty mattresses and, in the case of the one farthest from me, an old man or woman (it's hard to tell), crumpled.

The doctor looks up at us with surprise: foreign faces aren't common in these lanes, much less at his clinic. We have to talk fast, and Josh places him at ease by explaining that he's an American academic, a professor, somebody of repute. At that, the doctor, who obviously considers himself a person of erudition, bucks up and tells us that he's sixty years old, and has been serving this village since 1968. When he started, he explains, he and his colleagues were trained to treat the simple ailments of daily life; advanced diagnoses, much less treatments, were never expected of them. "In the sixties, seventies, and early eighties," he says, "most of the diseases around here were related to stomach problems, diarrhea, things related to diet and the water."

The poverty-related ailments disappeared just as soon as the county could afford to dig for deeper, better wells not contaminated by human and animal waste. Progress, however, has a price: those wells, he reminds us, were paid for by the plastic waste trade just up the street. "Since the eighties, high blood pressure has exploded," he explains. "In the past nobody had it. Now forty percent of the adults in this village have it. Back in the eighties, you'd only see it in people in their forties. In the nineties, we started seeing it in people over thirty. Now we're seeing it in people age twenty-eight and up. And it comes with pulmonary problems that restrict movement. People have it in their thirties so badly that they can't move anymore. They're paralyzed."

In the weeks following this visit I'll call a friend, a medical doctor, who tells me that the symptoms and the environment suggest that young villagers are developing pulmonary fibrosis and paralyzing strokes.

"Back in the seventies and eighties, you didn't die from high blood pressure," the village doctor adds. "Now you die from it. I'm sixty years old, and when I was a kid, I remember maybe one person who was so sick with it that he couldn't get out of bed [a likely stroke victim]. Now there are hundreds of people like that."

"What's the cause?" I ask.

He shrugs. "Pollution. It's one hundred percent pollution."

"Was it worth it?" I ask. "Was the cost to the environment and people's health worth the development of Wen'an?"

He shakes his head. "Health was better in the past. You knew what was wrong. But the sicknesses now, they'll kill you." He smiles at us. "Even I don't feel good. After you leave, I plan to go to the hospital, too."

CHAPTER 10

The Reincarnation Department

That scrapyard smell is strong, even when you're driving in an air-conditioned pickup. I'm seated next to Dave Stage, the affable manager of OmniSource's immense scrapyard in Fort Wayne, Indiana, and he's giving me a brief driving tour of the facility, along with my soon-to-be wife, Christine, in the jump seat. This is a first for me: I don't usually take anyone on reporting trips. But if we're going to marry, then Christine needs to know why I like all this scrap so much.

Dave slows to point out stacks of junked Coke and Pepsi vending machines. There must be hundreds. "Most of them still have fluorescent bulbs and refrigerator compressors that we have to pull out," he says. Both items contain hazardous materials that need to be recycled elsewhere. "After that, we shred them."

I look back at Christine. "In the automobile shredder."

She flashes me an annoyed look meant to convey that, yes, she knows. But I know she doesn't: until you've seen your first auto shredder, you can't know.

We drive parallel to bright, shiny piles of steel sheets from which shapes have been punched, and purchased, according to Dave, from a manufacturer of automobile clutches. "See over there," Dave says, and nods in the direction of a distant figure wearing a welding mask and wielding an acetylene torch with which he's cutting apart a bale of metal. "He's eighty years old. And he has one leg. Been coming here for years." I double-take: he does in fact have one leg. I look back at Christine: she's smiling from behind her sunglasses, getting into the spirit, I think.

Dave parks the truck at the office, and we step into a hot, yellow brass sun. All around me I hear the groan and crunch of scrap-processing equipment at work, turning ordinary objects into raw materials. "You'll need these." Dave hands us hard hats, safety glasses, and orange vests.

There are two piles of scrap steel in front of us, both roughly two stories tall. The one on the left is rusty in color, made up mostly of unrecognizable pieces of metal cut into short lengths that can be managed—and melted—by a steel mill. In industry parlance, that's "prepared steel."

The pile to the right, meanwhile, is a multicolored crush of everything from fences to old machine tools, scaffolding, pipes, a few bicycles, at least one swing set, and lots of shelving. This is known as "unprepared steel," meaning that it needs to be cut down to size and cleaned of anything that isn't steel before it can be sent off to a mill. Some of that work is done by hand and tool, and some of that work is done by dinosaur-like cranes outfitted with giant beaklike pincers that cut thick pieces of pipe like it's pasta.

There is also a third way to handle that steel, and it rises three stories above the scrapyard, casting a long shadow. From here, we can't see much except the carwide conveyor that's currently lifting a crushed auto body skyward. But from the side it looks like a contained carnival ride—it goes up, and then it goes down. At the top, steam rises, along with an otherworldly screech and groan—that's the sound of an automobile being reduced to fragments the size of my reporter's notebook. From here, at least, we can't see how.

Dave walks us around the scrap piles as two cranes swing back and forth, using their fingerlike grapples to unload two recently arrived trucks loaded with bed frames. When they're both empty, one of the cranes grabs a balled-up piece of chain-link fence and rubs the empty beds of both trucks, cleaning them out as if using a scouring pad. "Gets every last bit of metal out of there," Dave says.

Christine opens her bag and snaps a photo.

We're walking beside a wall of cars, six cars high and perhaps fifty cars long. They're a sad, deflated sight, emptied of engines, radiators, transmissions, wheels, tires, and anything else worth the trouble of extracting by hand. According to Dave cars arrive every day, dozens of them, and then bit by bit they're picked off by the cranes and deposited onto the hungry shredder conveyor. As we draw closer to the machine, the sound drowns thoughts, and Dave's voice rises. "We do one hundred and thirty tons per hour in the shredder," he yells in my ear.

"How many cars is that?" I yell in return.

"Seven to nine thousand cars per month. Depends on time of year, market."

OmniSource's shredder is really big, but what's most unusual about it might be that it's not unusual at all. North America is home to more than three hundred metal shredders (they're not all devoted to automobiles). Another five hundred, at least, are located in dozens of other countries from South Africa to Brazil, China to Sweden. The shredder stands as the singularly most important piece of recycling equipment ever developed. It is, among other things, the best and really only solution to managing the biggest source of consumer waste in the world today: the roughly 14 million American automobiles that are junked annually.

"Watch your step," Dave says as we reach the metal staircase that rises to the shredder's control booth. As we climb, I look back at Christine, and I notice her lips are held tight—perhaps protecting her mouth from the grime. But it's the noise that I notice. The higher we get, the closer we get to the point of shredding. From here, I can *feel* the individual screeches and screams of metal being crunched and pulverized.

At the top Dave opens a door, and the three of us step into unexpected air-conditioning. The door closes, and we're enveloped in relative quiet. "Welcome to the control room," Dave says. It's a bathroom-size space, staffed by Rob, a handsome half-shaved kid in his mid-twenties (I'd guess). He stands at the far corner of the booth with a window overlooking the mouth of the shredder, a monitor that displays an infrared picture of the heat being generated in the shredder itself (it looks like a television weather radar map), and a control panel with a joystick that he can use to manipulate the speed of material going into the system. "You know when a car is going through," he tells us. "It feels like an earthquake."

"Only then?" I ask.

I stare at the computer screen and its pulses of color. Somewhere in there, a car is being reduced to fragments by toddler-size hammers that weigh several hundred pounds each, spinning freely from the edges of a high-speed, love-seat-scale rotor. If they hit something too hard to break on first impact, they'll deflect all the way around and hit it again.

I turn around and look at the long sprawl of the scrapyard through the rear window, and—in the distance, over the trees—the handful of buildings that constitute Fort Wayne's modest skyline. Suddenly the booth shakes like it's been hit by a tank, and the air is filled with a deep, baritone groan. I look at Rob. "Car," he sighs, unbothered, his hand on the joystick as he watches a red blob on the infrared screen. On the other side of the glass, steam rises from the shredder's open mouth, the last breaths of something that once had the new car smell.

As the front end of that car is pulverized, the pieces are ejected through grates and run over magnets that separate the 80 percent of a car that's steel from everything else.

Thus, the steel in a car is harvested.

In 1969, seventy thousand automobiles and trucks were abandoned by their owners on the streets of New York City. Some of them leaked gas and oil; some of them provided habitat for rats and mosquitos; most were unsightly. And the problem was not exclusive to gritty mid-twentieth-century New York.

In 1970 the Institute of Scrap Iron and Steel, then the largest trade association representing the scrap industry, convened a conference on America's abandoned car problem. Frederick Uhlig, a senior engineer from General Motors, was one of the featured speakers, and when it came time for him to describe the scale of the abandoned car problem in America, he suggested that since 1955 Americans had abandoned between 9 and 40 million automobiles in fields, open bodies of water, and city streets. It was an absurd "estimate," reflective of the fact that the American automobile industry (much less its customers) had never taken much responsibility for the afterlife of the products that it placed on the road. It wasn't completely off the mark, either: the other experts at the same conference offered estimates on abandoned cars that ranged from 17 to 30 million.

Abandoned automobiles were not a new problem. In the 1920s, just a couple of decades into the American automobile era, Americans were retiring as many as 1 million cars per year. And there were a lot of cars to retire: the Ford Company manufactured 15 million Model Ts between

1905 and 1927 (without any recycling option for its customers). World War II slowed down the growth in auto buying, and forced Americans to repair what they'd otherwise retire for new, but by the early 1950s they'd returned to their car-buying and -abandoning ways. In 1951, the United State was home to an estimated 25,000 auto junkyards, and every last one was necessary: that year, 3.7 million cars and 600,000 light trucks were sold to scrapyards.

In general, the scrapyards were able to recycle those vehicles, with much of the work accomplished by hand. Workers, often in teams, would use axes and other hand tools to rip out the copper, the aluminum, the fabric, and the wood. Steel was the major leftover, easily marketed to American steel mills eager for raw materials to power midcentury American growth. After World War II some scrap companies developed giant shears to take care of the cutting work and keep up with the growing wave of old cars hitting the junkyards (the giant cutting machines, however, weren't nearly as good as human teams at pulling out the various nonsteel bits of a car). Tom McCarthy, writing in *Auto Mania*, a seminal history of the automobile's impact on the American environment, summarized the impact of the supply chain as follows: "The work of junkmen and scrap dealers remained the most significant step to reduce the environmental impact associated with the automobile's lifecyle before the 1960s. For their efforts, they received little positive recognition."

Such is the life of the American junk man!

Two very significant blows fundamentally altered the American auto-junking business in the mid-1950s. First, the price of American labor began to rise, making it increasingly difficult for U.S. scrapyards to pay teams of workers to tear down cars into their various components. Meanwhile, American steel mills began to upgrade their technologies, and by the early 1950s many were no longer interested in melting down old automobile bodies procured from scrapyards. The problem was copper: even a small amount—1 percent or so—when melted in a steel furnace will weaken the properties of steel. Scrapyard teams might have been good at stripping cars of their various metals, but extracting the last bit of copper was difficult—and expensive. By the mid-1950s, with many American steel mills no longer interested in buying cars from scrapyards, those same scrapyards stopped buying cars from Americans. The resulting cascade was predictable: Americans began abandoning their cars by the millions.

By 1960 the abandoned automobile problem was a massive environ-mental crisis. Across America, streams, creeks, and fields were polluted by cars and their leaking oil, gas, and other fluids; junkyards with stacks of old automobiles destroyed views of America's once-pristine country-side. Recycling, once practiced intensely, was no longer an option, forcing governments and business alike to seek alternatives. Those alternatives, in retrospect, seem absurd: when a massive earthquake struck Anchorage, Alaska, in 1964, destroying thousands of cars, residents disposed of them by dropping the vehicles off a 350-foot cliff; local Florida governments, overwhelmed by the hulks, started dumping them in the ocean out of a misplaced hope that they might form reefs. By the mid-1960s Lady Bird Johnson, wife of the thirty-sixth president, was so appalled by the nega-tive impact of mega-junkyards on the American countryside that she pushed highway "beautification" legislation that required—among other improvements—that by 1970 any junkyard located within a thou-sand feet of a federal interstate be concealed by a fence or removed.

It didn't work out. There were just too many cars, and they had to go somewhere; better the junkyard by the highway than in my front yard.

On February 10, 1970, President Richard Nixon told the U.S. Con-gress, "Few of America's eyesores are so unsightly as its millions of junked automobiles." More likely than not, Nixon was unaware that a solution to the problem was starting to grind away. It would take de-cades to catch up on the backlog, but the problem—more or less—was being solved by the first automobile shredders.

That Americans in the twenty-first century spend more time worry-ing about what happens to their unwanted cell phones than their junked cars—junked cars that weigh tens of millions of tons more than all of the e-scrap generated in the world every year—is no small testament to whom-ever decided that the best way to recycle a car is to obliterate it entirely.

The shredder booth shakes again. The air groans again. I lean to look out the window, but there's only steam. I glance back at Christine, on her toes looking past me and Rob, trying to see the actual destruction of an automobile. But it's impossible: the actual shredding takes place in-side an armored box.

"Seen enough?" Dave asks.

I want to say, Not really. But it's Dave's shredder, and Dave's tour, and so we follow him down the stairs and behind the shredder, where shiny pieces of steel crumpled like sheets of paper are speeding by on conveyors. You don't want to get too close, though: they're sharp, and uncomfortably hot to the touch. As we're walking to the end of the line, we pause beside a set of nine manhole-size steel discs strung together on a rotor like a child's hard candy bracelet, and hung between two even thicker discs. "That's the rotor we'll switch out," Dave says. Large pins protrude from the surfaces of the smaller discs (we can't see them from the side), and large triangular, super-hardened, several-hundred-pound steel hammers roughly the thickness of a man's thigh are hung from them. When the rotor spins, the hammers emerge from the spaces between the discs, and pound whatever gets in their way.

I'm tempted to tell Christine that—according to somebody I met the other day—a rotor like that can cost in the range of $500,000. But I let it go until later.

We're behind the shredder now, and it's actually relatively quiet back here. Above us, a thin conveyor lifts the stream of crumpled steel twenty feet into the air until it falls into a ten-foot-high pile, the pieces ringing like a wind chime in a breeze. It doesn't remain there long: a crane outfitted with a kitchen-table-size magnet is transferring the shredded metal, dip by magnetic dip, into a trailer bound for a steel mill.

For those who don't want to obtain their steelmaking materials from expensive, environmentally destructive iron ore mines, this is a windfall. Conservatively estimated, each ton of shredded metal loaded into that trailer is equal to 2,500 pounds of iron ore that won't be mined in northern Minnesota (or elsewhere), and 1400 pounds of coal that won't be dug up in Kentucky to power the steel furnace. To be sure, automobile shredders require ungodly amounts of power, but compared to the electricity necessary to run an iron mine, they're downright stingy. It's one reason, among many, that Steel Dynamics—one of North America's largest steelmakers—bought family-owned OmniSource for $1.1 billion in 2007. In 2013 the U.S. scrap industry processed 77 million tons of iron and steel, roughly half of which was shredded. And that shredded scrap metal, when remelted, accounted for roughly 30 percent of the new steel manufactured in the United States.

Christine takes the camera out of her bag, steps in front of us, and snaps a photo of the shredded steel streaming gently into a pile. Then she continues along, unbothered.

But later, when we're alone, she tells me: "That's the sexiest, most masculine machine I've ever seen. Only men could come up with something like that. Men."

Four months later, at six on a cold January morning outside El Paso, Texas, Scott Newell, CEO of the Shredder Company, supplier of more than 30 percent of the world's metal shredders, picks me up from my motel. The rough stubble on his face must be three or four days old, but on his taut-skinned face it makes him look vital. He may be seventy-two, but if I didn't know it, I'd say he's fifty-five. "What'd you think of that book I gave you?" he asks as we drive off in his Escalade.

Yesterday, while in his office briefly, I saw he had a copy of *Keynes-Hayek*, an intellectual history of the two great economists, on his desk. He wasn't done with it, but he insisted that I take it anyway.

"Deep," I say, not willing to admit that I fell asleep in my clothes after a full day of El Paso scrapyards.

Scott is of modest stature, and has a quick step—two qualities that, in combination, I associate with the fleet-footed academics I knew in college, and the fleet-footed scrap traders I knew as a child. Before I met Scott, and before I took Christine to OmniSource, I can't say that I'd ever thought about what kind of man would design a machine that eats cars whole. But now it occurs to me that it would have to be someone like Scott. And in fact it was: Scott's father, Alton Newell, didn't invent the shredder, but he definitely perfected it. Of the eight hundred or so shredders now in existence, more than half were built by, based on, or use ideas "borrowed from" (Scott's phrase) Alton Newell's designs. From the beginning, Scott was right there alongside his father, building the machines, refining the designs, and then—notably—using them at Newell family scrapyards in Texas.

The first Newell shredder was built for the family's flagship San Antonio yard in 1959. The second was built there, too, and today it sports a plaque naming it a National Engineering Landmark (so says the American Society of Mechanical Engineers). By the end of the decade, the Newells were building or licensing similar shredders for companies across the United States and around the world (Scott has very fond memories of visiting the Newell office in Rio de Janeiro). At first they competed with mega-shredders. But where those bigger shredders were favored by the biggest scrap companies and steel mills, who could afford them, the Newell shredders were affordable to smaller scrap companies.

As Scott explains it to me in the SUV, what made that shift possible was "The Insight," as had by his father. The mega-shredders that preceded the Newell shredder were crude rock crushers into which cars were literally dropped onto spinning hammers and then torn up in a process that I imagine was something like what would happen if I dropped a toy car into a blender. The system worked—cars were definitely reduced to fragments—but at a tremendous cost in power and maintenance. Alton Newell's insight was that the same result—a shredded car—could be accomplished in a smaller machine, requiring less power and investment, if you just slowly fed the car through two rollers, inch by inch into the spinning hammers, and let them nibble away at it. To be sure, there were other refinements along the way, but it was the side feed roller, The Insight, that made the shredder affordable to smaller scrapyards and cleaned up the mess of cars polluting the American countryside.

It starts to feel like dawn as we drive through the small town of Canutillo, past piles of under-construction windmill towers set on their sides like so many monumental drinking straws. "It's a subsidized business," Scott says of wind power, with the contempt of an entrepreneur who's never known such a luxury. "Can't stand on its own." We're on a slight decline, and as he brakes for a stop sign I can see the ghostly lights of the Shredder Company, burning all night, spread over eighteen acres.

Pulling up the driveway, we pass a flatbed semi trailer loaded with a shredder rotor roughly the size of an aged tree trunk. "That one's going to China," he tells me as we pass it. "No, Ecuador," he corrects himself. "It's going to Ecuador."

He parks the SUV next to a small single-story building beside several much larger warehouses, and we step out. "One of my former employees is now a missionary in Ecuador," he explains. "And he helped to arrange that deal. We have *other* pieces going to China."

It's so quiet standing out here, so remote. I look down the long, deep valley that opens into Mexico, but there's nothing to it but the occasional twinkling light. Then I notice a mechanical humming that starts to fill the early morning. It's not deep, more a tenor, and it sounds busy. "That's the furnace," Scott explains gently.

We walk through a loading bay into a sports-arena-size industrial space that flashes orange for a moment, then reverts to steel brown and gray. High above us, lights cast triangular halos that drop from above onto men in hard hats, miniaturized by the space, scurrying around piles of sand and suitcase-size metal frames. I stop beside one of the frames: it holds a mold for three bell-shaped shredder hammers. Opposite are more molds, but those are covered with black sand and steaming. I step closer for a look, but Scott warns me to be careful: they're freshly poured and hot.

I notice, as we venture into the foundry, a large metal caldron and, next to it, a long steel box, roughly the size and shape of a cattle trough, filled with big hunks of round scrap metal. "Those are returned hammers," Scott tells me.

"Used hammers?" I ask.

"Yeah. We take a lot back once they're finished. Melt them into new ones. We also take rock crusher parts."

The sharp bell-shaped metal is now worn into round nubs that merely look heavy—no longer dangerous. They're spent. According to Scott, each ton of steel that's shredded in a shredder wears away one kilo—or 2.2 pounds—of cast steel, mostly hammers, but also other parts. A simpler way to think about it is this: every one-ton Volkswagen Beetle rips away 2.2 pounds of shredder—the car's last 2.2 pounds of flesh, if you will—as it disintegrates. Eventually, the hammers wear down so far that they need to be replaced.

I stroll past the box of worn hammers and come upon several trough-size boxes filled with what are unmistakably fragments of shredded steel of the sort Christine and I saw pouring off conveyors at OmniSource a few months ago. It's mangled and a touch rusty, and—away from the shredder—it's nothing special. It's just metal, sprinkled with white flecks that Scott tells me are limestone (in the furnace, the limestone combines with—and removes—impurities in the metal). Later in the morning the box will be lifted up to the furnace, melted, and then poured into new castings—hammers, perhaps. "The steel you see in there comes from our own shredder across town," he tells me. It's one of the earliest shredders, in fact, dating back to the 1960s, and Scott proudly keeps it running. "It's like the family ax," Scott jokes. "You change the handle. You change the blade. But it's still the same old family ax."

I glance around the capacious room, its grays and browns uninterrupted, and it occurs to me that this is, in a sense, Green Heaven, where recycling equipment is recycled into new recycling equipment, shredded cars transformed into the means by which other cars will be shredded. It's a solution to a problem that most Americans don't even realize is a problem: how to get rid of their cars in a manner that reuses as much of the car as possible.

Alton Newell was born in 1913 to a family of poor migrant sharecroppers who bounced between California and Oklahoma, living a hard life that resembled, in many respects, that of the Joads of *Grapes of Wrath*. At one point they traveled in a caravan of three cars, sharing meals and campsites while working in fruit groves. When the cars broke down, they fixed them using whatever parts they could find—and the place to find parts in those days was junkyards. It was perhaps natural, then, that in his teens Newell took a job at a scrapyard that specialized in "wrecking" cars in Santa Ana, California.

The late 1920s were a promising time to get into automobile recycling. Americans were just beginning to realize that they might have a problem disposing of that first generation of automobiles that had driven their last miles. Thousands of auto-wrecking businesses sprang up across the United States, mostly outside the big cities, and accepted cars for free or for a small fee. In very rare cases, they might pay for them. It was easy money: most cars had at least a few usable parts that could be pulled off, reconditioned, and resold. The hard money, the early auto wreckers learned quickly, was the 80 percent of the car that was made of steel in those days.

Theoretically, at least, steel should be recyclable. But when it comes to automobiles, that's not always the case. From the perspective of a steel mill, an automobile isn't just steel—rather, it's a complicated bundle of upholstery, rubber, glass, a range of nonferrous (that is, not steel or iron) metals like aluminum, zinc, and copper—*and* steel. Many of those materials—especially copper—when melted with steel in a furnace alter the properties of the steel so that it can no longer be used. Of course, it's possible to strip out most of the contaminants from the steel by hand, but doing so takes hours and requires a good reason. Most people with enough to eat don't have one.

Newell Recycling of Atlanta, now run by Alton Newell's daughter, claims on its website that at one time Alton Newell could disassemble an entire automobile on his own in a mere ten hours, using an ax. Many auto wreckers—including Alton Newell, who acquired his own business in 1938—discovered that at least some of their problems could be solved with a box of matches and some gasoline. Upholstery, floorboards, and carpeting are eminently combustible, and once burned away leave a shell that is much easier to strip of whatever doesn't burn or melt. It's by no means a perfect method, mostly because nobody enjoys living downwind from a burning car (just try it), but also because steel mills don't exactly relish charred scrap, either.

Meanwhile, up in Detroit, Henry Ford too saw commercial opportunities in abandoned cars. But rather than set up a large-scale burning operation, Ford tried to replicate the success he had with assembly lines by establishing a large-scale *disassembly* line. The idea was that economies of scale would take care of the profitability issue. So, for example, the seat stuffing from one Ford Model T might be a worthless nuisance, but the stuffing from hundreds might be something that could be sold, or reused in new Model Ts. Ford's 1934 biographer, Robert Graves, referred to the operation as the "Reincarnation Department," and during its short life it was apparently a sight to behold: hundreds of men deconstructed hundreds of automobiles per day into their components. There was only one problem: the Reincarnation Department lost lots and lots of money. The economies of scale were real, but so was the cost involved in paying hundreds of men to achieve them. And so, over the course of the 1930s, the Reincarnation Department was slowly laid to rest, piece by piece, line by line, as the business was reclaimed by the scrapyards.

Auto salvage was a business waiting for an innovation that removed impurities from steel for the lowest possible cost. Alas, the most popular innovation was to move the incineration of cars into contained "pit incinerators." These were basically gas-fired automobile crematoriums dug beneath ground and encased in concrete. Cars were wheeled into them on tracks, torched, and then pulled back up, free of everything but the metal that wouldn't melt. It was a fine way to take care of the problem, so long as nobody noticed the pollution. But smoke is black, and in

the mid-1960s the National Air Pollution Control Administration collected and published data suggesting that 5 percent of all air pollution in
the United States was caused by automobile incineration. Local governments that had once worried about abandoned cars now shifted their
attention to shutting down the solution.

Who could blame them? I've traveled for ten years in some of the
harshest scrap-metal processing zones in Asia, and I've seen some of the
worst recycling practices that an entrepreneur could possibly invent, and
yet I must concede that I've rarely seen anything quite so awful as an incinerator devoted to the (mostly) unfiltered cremation of dozens of cars
per day. The stink must have been awful; the environmental damage done,
unspeakable.

But shutting down the incinerators didn't solve the abandoned car
problem. In 1965 Americans tossed 9.6 million automobiles, only around
1 million of which were actually recycled. The results were ugly: according to data reported by the Institute of Scrap Iron and Steel, at least 20
million abandoned cars were scattered around the United States in 1970.
Nobody was interested in buying them. Under such circumstances, I
suppose, it only made sense to dump your old car in a creek.

Scott Newell takes me for red tamales at a small Mexican restaurant
near the Shredder Company. The waitresses greet him by name, and he
orders with little more than a glance at the menu. Then we sit. "You ask,
Why do you shred cars?" he says to me. "It was because a car had upholstery seats, rubber, brass . . . we had to burn them in the open."

In 1955 Sammy Proler of Proler Steel in Houston, Texas, was thinking
along these same lines. Proler had a big problem: he owned 40,000 tons
of junk automobiles that he could not burn, and could not afford to strip
by hand. But he absolutely had to unload them. One afternoon he boarded
a flight from Salt Lake City to Omaha, and several hours and four screwdrivers later, he had a solution: shred them and then run the fragments
over magnets to reclaim the steel.

The idea wasn't revolutionary, nor was it as crazy as it might sound
now. Scrap companies had been shredding tin cans since 1928 (it was
easier to chemically remove tin plating if the cans were shredded sheets
rather than whole cylinders), and those same shredders had progres-
sively been modified to shred bigger and bigger items, including thin

car parts, like doors. Sammy Proler's shredder—formally named the Prolerizer—debuted in 1958 at nearly one-fifth of a mile in length, powered by motors pulled from U.S. Navy warships. It was so powerful that all one needed to do was drop a car onto its spinning hammers and you'd have near instant scrap. There was no smoke, and the expense, though steep, was cheap compared to trying to accomplish the same task by hand.

Meanwhile, in San Antonio, Alton Newell had become the owner of the town's biggest scrap company. And like the Prolers, he was involved in wrecking cars; but unlike them, he didn't have 40,000 tons of junkers sitting around. Still, Newell also had experience shredding tin cans, and like Proler, he saw the potential in shredding bigger and bigger pieces of junk.

Over lunch, Scott Newell tells me that his father had started thinking of shredding cars before the Prolerizer came online. In fact, the Newells were already shredding car parts, and firmly believed that a whole car could be shredded too. Then the news of the Prolerizer arrived, and Newell had reason to try: "As soon as we *saw* that that could be done [shred a car] . . . my dad built a bigger machine, one wide enough to take a whole car." That was 1959. The Prolers, Scott Newell readily concedes, shredded first. But his family, he not so subtly intimates, shredded better.

Later, Scott takes me back to the Shredder Company and leads me into a conference room where a wooden scale model of a shredder sits in a corner. It's the size of a large microwave oven—a microwave oven with a conveyor that runs out the back end.

He grabs a handle that lifts the entire left side of the shredder, including The Insight: that side feed roller that slowly delivers a car to the hammers (instead of dropping it from the top). But that's not what Scott wants to show me. He's lifting up the smaller box that contains the rotor and hammers. "The housing opens this way," he says, fully revealing the segmented rotor—the same shape as the one in Indiana, though thousands of times lighter. There's a handle that's connected to it, and Scott can't help himself; he starts turning it like a toy rolling pin, and as he does, the force of the spin pushes the little hammers out of their slots between the discs.

In reality, the rotor spins counterclockwise onto the hood of a car as it enters the chamber. The shredded material then spins around the chamber and, once it's been reduced to a size small enough, is propelled upward through a grate. More or less, that's the same design used in the first Newell shredder. The differences are subtle, learned and developed "through fifty years of mistakes," Scott concedes. "Many made by me."

As Scott reassembles the model, he suddenly shifts the subject. "I have trouble when I take this to a Chinese exhibit. All of our potential customers are over there." He mimes taking photos of the model, and starts to laugh. "Making blueprints. Blueprints with their cameras."

But as Scott knows better than anyone, the shredder isn't a secret anymore. It's a fifty-year-old solution to a problem that confronts most developing countries at some point: too many cars and too few people willing to work for the low wages necessary to disassemble them by hand for recycling.

Early summer 2008, the global economy is flush with cash but ready to collapse. At the direction of *Scrap* magazine, I travel to Bangkok to figure out why a small country like Thailand has become—in very short order—a major importer of American shredded scrap metal. In a sense, we already know before I fly down from Shanghai: Southeast Asia, in the shadow of China, is undergoing a construction boom that requires more and more steel to build more and more buildings and cars, and Americans generate an excess of scrap metal.

Over the course of several days, I visit a handful of major steel mills in the Bangkok area. They are flush with scrap, much of it imported. At GJ Steel, one of Thailand's biggest steel companies, a manager tells me that they have 120,000 tons of steel just lying on the ground. Then, after a beat, he asks me: "Is that a lot?"

It is, I assure him. That's equivalent to the weight of a very large oil tanker, or perhaps 15 percent of the total American steel scrap exported to Thailand that year. We are in the mill, watching as a stream of imported shredded scrap as wide as a creek flows into one of the largest furnaces in Asia. That furnace looks like a science fiction nightmare: a flying saucer covered in dust and piping and shooting sparks. The heat is unbearable—we can't draw too close—and I'm led instead into a control room where a group of technicians is looking at a monitor that displays the precise recipe for the steel that they'll be making over the next few minutes.

- 20 ton—PIG IRON (an iron ore product used in steelmaking)
- 30 ton—HMS MIX (iron and steel scrap)
- 20 ton—BUNDLE NO. 1 (steel sheets in a bundle)
- 60 ton—BUSHELING (free, clean steel scrap)
- 60 ton—SHREDDED IMP (IMP, as in imported)

Why imported? Because Americans and Europeans don't use their metal goods—like cars—long enough to let them rust, they generally produce higher-quality scrap metal, including shredded metal, that melts better than local scrap. But even if Thai shredded scrap were equal in quality to American shredded scrap, there simply isn't enough of it in the market to fully supply all of the steel mills rushing to meet these boom times.

I leave the control room and walk around a scaffolding to where new sheets of steel are flowing out of machines, silver and clean, until they're rolled up into giant shiny tubes and left to cool. In coming days they'll be shipped to manufacturers of cars, washing machines, and other makers of middle-class comforts across Thailand. It's amazing to me, watching that reincarnated steel flow out the door, into the lives of people who will never know that, perhaps, an American once took his wife on dates in it. When I was a kid, I just assumed the old cars we sold to the local shredder would make their way back to our yard, eventually. That cycle is outdated.

It bothers me.

Later that night, just after nine, I'm in the back seat of a taxi in Bangkok with Randy Goodman, then the director of international marketing and logistics for OmniSource Corporation (and now an executive VP with a smaller, family-owned business in Kentucky). We're both in town for a conference, and we've just finished dinner. As we head back to the hotel, I tell him how odd it was to see so much imported shredded scrap on the outskirts of Bangkok.

"They can't get it at home," he says. "So they'll come looking to us for it. What do you think I'm doing here?"

Randy is a big man, and he's sweating in a Hawaiian shirt, suffering through a sultry traffic jam. His BlackBerry buzzes. "This is Randy."

It's the home office. I turn to look out the window at a city draped in condensation and heat. There's a skinny man with a pushcart walking along the sidewalk, carrying what looks to me like pieces of spindly rebar pulled from a demolition site. The price of scrap steel is just as flush as the economy that day, and I suspect that the peddler will make a few bucks. Earlier in the morning, I had several conversations with traders at the conference who told me, candidly, that they expected prices to break through the once unthinkable $1,000-per-metric-ton barrier, and never turning back. It's unbelievable to me: when I was a teenager, we sometimes *charged* to take steel scrap, it was worth so little. That was before China, though; that was before Asia needed metal. But now, at what turns out to be the last moments of the global economic expansion, anything steel is worth selling. Back in China, manhole covers are disappearing from streets in the middle of the night.

"They're like rollers, right?" Randy asks, in his deep voice. "With rubber on them. Okay. Yeah, let me see what I can do. Send me some photos." He says good-bye and hangs up the phone. "They're looking for a price on these giant steel rollers that are, like, encased in rubber."

"What're they for?"

"Used at fabric mills," he says.

"Who buys them?"

"In this market? Anybody who buys steel. The price is high enough they'll figure out the rubber. But in another market, probably not."

Steel wrapped in thick rubber is not, generally, the sort of scrap that buyers want except when they're desperate. And in the summer of 2008, with scrap prices reaching new heights, panic buyers take any kind of steel they can find, for fear that prices will be higher in the morning—which in fact they usually are.

I forget about this taxi ride until three years later, summer of 2012, when I find myself and Christine in the office of Dave Stage, Omni-Source's Fort Wayne scrapyard manager. We've just finished our shredder tour, and now we're chatting about how hot the scrap markets were right before everything crashed in the fall of 2008. I tell Dave about that night with Randy.

"I know what those rollers were," he says with a nod. "I've seen those. The market was crazy in 2008."

"Really?"

He laughs. "Sure, the summer of 2008 we had a line to drop scrap at the shredder that stretched all the way to downtown Fort Wayne. Two miles."

This piques Christine's attention. "Really?"

"Seven hours long. That's what happens when the prices get real high. People start pulling stuff out of the garage, out of the woods, out of the fields, that they wouldn't touch otherwise."

A few months earlier I sat down with Dennis Reno, Sr., a senior vice president at Huron Valley Steel, the company founded by Leonard Fritz outside of Detroit. Huron Valley's shredder happened to be Detroit's first shredder, and Reno told me that the pent-up demand for a place to take—and be paid for—junked cars was so strong that the company had traffic jams going into the yard in the late 1960s when it first started running. Even so, according to Reno it wasn't until the early 1990s that the pent-up stockpile of abandoned automobiles began to be whittled down, and not until the 2008 run-up in steel prices was the U.S. scrap industry finally caught up and just recycling what was entering the junkyards. "We've been pretty current since then."

Sitting across from Dave, I ask if he agrees. Was 2008 the year that the last of the scrap cars came out of the woods?

"Tell you what. One day in the summer of 2008 we had an old farm tractor come into the yard for the shredder. It had a tree sticking out of it, with a trunk with a circumference the size of a baseball. That's how long it'd been sitting there. So yeah, I think it probably was 2008."

"And the lines were seven hours long?"

He laughs. "We had two, three cops on the corner. A guy set up a sandwich stand. We'd call up Subway, Pizza Hut, and order a hundred and twenty pan pizzas. Truck drivers were setting up barbecue grills in the yard."

I look at Christine, and she's laughing with Dave. It's funny, for sure. But it's also astonishing, when you think about it: eighty years after the first Model T rolled off Henry Ford's assembly line, Americans finally managed to clean up a backlog of junked vehicles—and they did so in part because steel mills in Bangkok needed raw materials to make new cars and refrigerators for people in Southeast Asia.

The end of the automobile backlog should've been occasion for a television documentary, a presidential speech, and perhaps a commemorative postage stamp. Instead, there were daily orders of pan pizzas and barbecues in the long shadow of one of North America's biggest Newell shredders. I just wish I'd been invited.

Instead, I was in China—the fastest-growing market in the world for shredders. Scott Newell has been busy here for nearly a decade, delivering some of the biggest shredders ever built to Chinese scrapyards and steel mills convinced that it's better to overbuy than underbuy. Someday, they're betting, China will shred more than the United States.

It's after midnight, midwinter 2010, and I'm drunk in the back seat of a car headed to a new shredder on the outskirts of Lianyungang, 270 miles north of Shanghai. Outside there's little to see but the miles and miles of evenly placed streetlamps that the local government has erected to illuminate this barren stretch of road. Inside, in the back seat beside me, there's Frank Huang, a Beijing-based sales representative for Metso Lindemann, a major Finnish supplier of equipment to the recycling industry. Frank, too, is drunk, and like me, he's unhappy about that fact. "This is what I have to do all of the time," he grumbles. "It's the way of doing business in China. If I don't drink, they think I'm unfriendly."

Only thirty minutes ago we were seated at a round table with managers from Armco Renewable Metals, the owner of the shredder we're en route to see. At the beginning of the meal, it must be said, the managers were cool to the idea of taking a foreign journalist to see their new multimillion-dollar piece of equipment. Who can blame them? Despite deep connections to government agencies and officials at both the local and national level, Frank tells me, Armco has had trouble obtaining enough electricity to run its new shredder. As a result, these days it runs only twice per week, and only after midnight. I wouldn't want to tell that to a journalist, either.

As the meal progressed, and I accepted every "bottoms up" toast of high-proof grain alcohol, the Armco management team appeared to recognize a kindred scrap-metal spirit. And it's true: I *am* a kindred spirit, full of tales of my family's own scrapyard and the difficulty of running a shredder in the United States—not that we had one, but by the end of the meal I think everyone believed that we did. By the end of the second bottle, I was feeling nauseous, and the Armco team was feeling like I was just the sort of foreigner who *should* see the shredder.

Joining us at the meal was Song Yanzhao, deputy general manager of the Shandong Yuxi Group, a big-time steel recycling company in Shandong Province founded by his father. He's seated in the front passenger seat. By the look of his baby face, he's maybe twenty-two, but by the look in his eyes he's losing his youth to the cold cynicism characteristic of anyone who grows too close to China's politically sensitive state-run steel industry. Earlier, before the drinking, he told me that his family started out simply, cleaning metal using hand tools and torch cutters, then cutting it into sizes that could be accepted by China's steel mills. They were successful: by 2005, using hand tools, cutting torches, and large alligator-like shears, they were able to process between two and three thousand tons per month—roughly, the weight of ten to fifteen blue whales—collected within a few kilometers of their scrapyard.

That was pretty good, but the unfulfilled opportunity was even bigger. By 2005 China's was the fastest-growing major economy in the world, and its people and companies were beginning to throw away steel—buildings, cars, signs—at a pace that simply couldn't be processed by manpower alone. If a company was going to clean up that growing volume of steel, it needed horsepower—specifically, a very modest 1,000-horsepower Chinese-built metal shredder. Song tells me Shandong Yuxi's shredder breaks down frequently, but no matter: in the space of a few years, the company has gone from scrapping the equivalent of ten local blue whales per month to scrapping fifty blue whales gathered in a collecting zone that extends 115 miles from their headquarters. Without the shredder, they'd be stuck at ten to twenty blue whales—and responsible for many more employees than they are now.

Still, Song knows there are much bigger opportunities in the company's future. In 2009 China surpassed the United States to become the world's largest automobile market, with 18 million vehicles sold. Already plenty of older vehicles are being retired, but the bigger wave is due at the end of the decade, when the Chinese start junking the decade-old cars that constituted this latest boom. Here's the problem: labor is still cheap in China, but even if it were cheap enough to allow for affordable hand dismantling of cars, there simply won't be enough human hands to dismantle the tens of millions of Chinese cars that will to need to be recycled over the next few years. China's coming car problem isn't just a resource recovery issue, it's a manpower issue. Without shredders, and lots of them, they'll have cars stacked up from Beijing to Shanghai.

A view of Qingyuan, population 3.7 million, from Homer Lai's apartment. In 1980 Qingyuan was a modest farming community.

On a typical morning Homer's scrapyard office is crowded with friends and relatives talking business. In this 2009 image, Johnson Zeng is far left and Homer is next to him, head turned, in a baseball cap.

A warehouse at Qingyuan Jintian Enterprise Company, the largest copper recycler in Qingyuan and a major destination for the scrap metal that Johnson Zeng sells through Homer Lai. Workers run cables through strippers.

Cables too small to be run through strippers are processed in chopping machines like those utilized at Raymond Li's Christmas tree light factory—but bigger.

A 2009 view of downtown Wen'an County, then the center of northern China's scrap plastics recycling trade and the most polluted place I have ever visited.

At a characteristic Wen'an plastics recycling factory, workers wash shredded plastics in caustic cleaning agents without safety protection.

Workers strip stickers from fruit baskets imported into Wen'an from Thailand.

Once the baskets are clean of stickers, a husband-and-wife team shreds and washes them in preparation for remelting.

In 1970 General Motors estimated that Americans had abandoned as many as 40 million automobiles in public places across the United States over the previous fifteen years. Bruce McAlister, a photographer with the Environmental Protection Agency's DOCUMERICA Project, photographed the damage during the early 1970s, including (above) a station wagon abandoned in Colorado's Clear Creek Canyon, and (below) cars abandoned in water near Salt Lake. *Images courtesy of the National Archives and Records Administration.*

Stacks of cars awaiting the shredder at OmniSource's Fort Wayne scrapyard in 2011. The company can shred thousands of cars per month. *Image courtesy of Christine Tan.*

A metal shredder in Lynchburg, Virginia. Automobiles and other scrap rise up the conveyor to the left, then are fed into the shredder box via the side feed roller, located at the lower right. *Image courtesy of Kent Kiser and* Scrap *magazine.*

One of two metal shredders at Toyota Metal in Handa, Japan, used to destroy and recycle test vehicles.

An engineer at Toyota Metal in Handa stands beside a used and new shredder hammer. Roughly two pounds of hammer steel are worn away with every ton of automobile shredded.

A close-up of shredded automobile steel at GJ Steel outside of Bangkok, Thailand.

Shredded imported scrap conveyed into GJ Steel's furnace, the largest steel scrap–melting furnace in Thailand at the time this photo was taken in 2008.

Steel scrap is melted at temperatures that can exceed 2700 degrees Fahrenheit (1500 degrees Celsius).

In a good month, the company can manufacture 100,000 metric tons of steel for sale to Thailand's automobile and appliance manufacturing sectors.

A used computer market in Shanghai. Many of the components used in these devices are imported as so-called e-waste.

Mobile phone parts at a warehouse in Guiyu, China. The large number of identical parts suggests that they are factory rejects, overruns, or leftovers from a repair program. Samsung-branded bags are scattered throughout the pile. Parts like these are sources of reusable components for China's secondhand electronics industry.

A display case filled with used computer chips at the Guiyu Electrical Components Market. Everything sold at the market is recovered from e-waste.

At the urging of a companion, I posed boldly in the aisles of the Guiyu market. My smile is more nervous than pleased—minutes later, my companion was told that we'd best leave town soon.

A 2002 image of the automobile scrap sorting warehouse at the Sigma Group's Shanghai facility.

In 2005 Sigma moved into a new Shanghai factory that employs as many as eight hundred hand-sorters.

Workers at Sigma Group sort imported shredded automobile scrap in 2002. The aluminum will be melted into new aluminum and likely exported to Japanese car makers.

After a car has been shredded and the steel removed, the leftovers are between 3 percent and 95 percent metal. This pile of Shredded Non-Ferrous Metal—or SNF—is at Huron Valley Steel Corporation in Belleville, Michigan.

A washing machine–size box of U.S. coins recovered from SNF at Huron Valley Steel Corporation. The average American car is shredded with $1.65 of change inside.

Small fragments of imported shredded automobile scrap are sorted at Junlong Metal Recycling in Foshan, China.

Experienced sorters can earn in excess of $500 per month and are regularly poached by other scrapyards.

A disassembled iPhone 4S. The rare earths and glass used in the screen and faceplate (to the upper right) cannot be recycled. *Image courtesy of iFixit.*

The battery in the 2012 Mac Book Pro is glued to the case, making it nearly impossible to replace or remove safely for recycling. *Image courtesy of iFixit.*

Old Chinese televisions await recycling at Hunan Vary in Miluo, China.

Hunan Vary's workers dismantle the televisions into individual components for recycling.

That's why Song is in the car: his father has entrusted him with finding a high-quality foreign-made shredder to handle the coming wave,
and he's asked Frank to show him a Metso machine in operation. Song
concedes that his family might be early to the automobile-shredding
business, but when that wave of cars washes across Shandong Province,
they want to be ready.

Ahead, I can see orange halos cast by sodium lights in the crisp, polluted winter air, and the silhouettes left by several two-story hills of scrap
metal. "That's Armco," Frank tells me. Our driver parks in the empty lot
beside the darkened two-story headquarters and a car with two inebriated Armco managers who are waiting for us. As we step outside, the
breeze cuts through my thin coat, sobering me up. But we're all drunk, no
doubt about it; anyone truly sober would slip into the headquarters in
search of coat and gloves. Instead we walk toward the sodium lights and
the distant, thin groan of a shredder whose control room is just peeking
over the hills of scrap it's been commissioned to consume.

I drop back a few steps and follow the small group into a two-lane
valley cut between hills of gangly scrap metal. It's ugly stuff, unlike anything I've ever seen in years of scrapyards: twisted, turned, and haunted
like broken bones. I pause to get a closer look, and realize that the maze
of metal is actually a pile of bicycle and motorcycle frames stripped of
their tires, chains, and motors. They're steel, rusty, bent, and bony, piled
atop junk generated by China's first wave of consumerism now gone to
waste. Pieces of indistinct sheet metal stick out of the thorny mix, as
well as metal frames, scaffolding, a metal closet, some signs, lots of
springs, fuel tanks, the occasional bicycle chain, and even a few car
doors. But it's the stripped motorcycle frames that get me. In the United
States nobody would bother to do that. The labor is too expensive. I
wonder: Is it cheap enough to do to cars, as well? Why would they need
a shredder?

Frank turns to me. "They plan to import American cars and start
shredding them, too."

"American cars?"

"There aren't enough Chinese cars yet." Walking together, we emerge
from behind the piles, and the shredder looms above us, three stories
high. Two orange cranes work beneath the sodium lights, grasping motorcycle frames in clawlike grapples and adding them to the mix moving up the conveyor.

"I'm a little surprised people strip down motorcycles," I say to Frank. "It's kind of expensive."

"Nobody will do it if they can't make money," he assures me.

We're at the base of the shredder, now, and when I look up there's The Insight, Alton Newell's feed roller, hidden behind a steel shield and a giant chain to turn it. That's no accident: it's manufactured by the Texas Shredder division of Metso Lindemann, and Texas Shredder was founded by individuals with direct links to Alton Newell and his inventions. The familiar sound of metal being pulverized, the grind and crunch, is loud, but it's not overwhelming. Somehow the night swallows most everything.

We climb the stairs on the back side of this 4,000-horsepower monster and stop on a landing where workers watch shredded pieces of metal shoot by on two conveyors. I look back toward the shredder housing where the hammers swing, but I only see steel and steam. I look in the opposite direction and see a giant, towering pile of shredded steel. "Where's it go?" I ask Frank.

He names an enormous, state-owned company in north China that makes steel products used in construction. So long as China's construction boom continues, then the demand for scrap steel to melt in furnaces for things like pipes, girders, and doors will continue. "They'll load it on barges and ship it up to their door," Frank adds.

Suddenly I feel dizzy and step back from the conveyor. For the first time in, I don't know, five minutes, I remember how drunk I am, and how ill advised it is to climb around a shredder—or a conveyor—when I'm in no position to pass a Breathalyzer. If my editor at *Scrap* (which publishes a scrapyard safety column) knew about this, he'd have a fit. I check my watch: it's nearly 1:00 A.M.

"Adam!"

Frank, who can clearly handle his liquor better than I can, is waving at me from the stairs, and I follow him into a stairwell that leads up to the control room atop the shredder. The exertion makes me want to vomit. But I can't, because now I'm in the tiny control room, and it's crowded up here.

A big window overlooks the tangled motorcycle frames as they drop into the shredder. Seated in front of it, on a raised platform that gives him a better view, is a man in his early thirties, perched in a chair that reminds me of the captain's seat on the deck of the Starship *Enterprise*. His hands hang loosely over two joysticks that he uses to control the flow of the scrap, while his eyes move between the conveyor and a computer monitor that gives him a detailed readout of what's happening inside the machine. In the darkness, it feels very sci-fi, as if I'm dropping in on the future.

My eyes, though, are drawn to that river of metal, rising from the concrete yard, passing briefly by the window, and then descending into a maw that emits only steam. Beyond it I can see the distant outlines of the road, and Lianyungang's perfect streetlights. South of here, in a town set on the Yangtze River, is one of the world's largest metal shredders. Scott Newell, son of Alton Newell, built and installed that one, and—I've checked—it too runs through the night, its hammers ripping up whatever steel the middle-class burghers of nearby Shanghai no longer want. It's a long way from Texas, a long way from sharecropping. But I don't think Alton Newell would be surprised by any of this. After all, shredding cars sure beats ten hours spent chopping up a car with a pickax. Just ask the guy in the chair with the joysticks in his hands which he'd rather be doing.

CHAPTER 11

The Golden Ingot

Early August 2010, and I've taken a subway to the north side of Shanghai with a clear plastic bag that holds five old cell phones that I plan to sell at a used electronics market. The old Motorola has been in the back of a drawer since 2004; the Nokia was beneath a pile of papers on the piano since early 2009 (I think). In the United States, they're junk-drawer fodder, unwanted by a culture (and me) fixated on the next upgrade. For the environmentally minded, they're something much worse: "e-waste," the catch-all term for old electronic devices that nobody wants anymore.

But here in China, my old phones aren't e-waste. Rather, they're a low-cost means to provide wireless communication to some of the hundreds of millions who can't afford or don't want to pay for a new phone.

At least, that's what I assume in August 2010.

I cross the street to the long, two-story used electronics building; it's covered in slightly out-of-date posters for computer hardware and, below, outdoor booths where young men with spiky hair sell beat-up desktop computers and monitors. The entries lack doors, instead using the long, clear plastic strips with which meatpacking houses keep in the cold as butchers move in and out with cuts of meat on forklifts.

Inside, the ceilings are low, and rows of fluorescent lights stretch the two-block length of the space. Banks of glass display cases contain computer hard drives, CD-ROM drives, keyboards, memory, and motherboards; atop them are laptops—used, refurbished, or entirely rebuilt, their screens wrapped in plastic to protect them from the dust. Everything is a little dingy, a little down at heel, and a lot like an American thrift shop. But never mind the atmosphere, it's the price that's right: for as little as $75 you can go home with a laptop that'll run a Web browser and Word.

I don't see many people over the age of thirty around here. Everyone seems to be college-age or younger, and they spend a lot of time chatting softly with companions, their faces tense with potential expense. It's where I'd come, too, if I were short of cash, and in need of computing power.

I ride the escalator to the second floor. Up here, the display cases run the length of the room, and contain phones in addition to the computers, hard drives, mice, and keyboards downstairs. The laptops are even more crowded, and cheaper, atop the displays, but somehow they look newer. The monitors are brighter; the cases are cleaner and polished. Below are phones, and those too look new; on first view, hardly any are as old as the ones in my bag. I lean over the cloudy glass in search of lost twins of my old phones, but as I search I'm distracted by the expansive odor of hot melted plastic.

It's coming from the right, where another spiky-haired kid is crouched over a desk, his right hand grasping a soldering iron that he's applying to the guts of a wide-open laptop. The shelves above him are filled with dozens of laptop cases lined up like books, with the spines turned out, and haphazard stacks of hard drives, CDs, and plastic bins filled with screws, bolts, and connectors. His workbench is even more of a mess. Pliers and voltage meters share space with even more hard drives, even more CDs, and wild tangles of cables mixed with cell phones. Still, he clearly knows what he's doing: newish-looking laptops and phones are for sale in the display case.

According to research done by Willie Cade, an American electronics refurbisher who advises the UN's Environmental Program, 25 percent of the hard drives sent for recycling and refurbishment in the United States have been used less than 500 hours. For all intents and purposes, those are new hard drives that can be used for hundreds of hours more.

But who in the United States will take the trouble to reuse them? Almost nobody, which is one very important reason—among many—that hard drives and other electronic wastes from the developed world move to the developing world. Since the mid-1980s, at least, China has been home to a thriving business devoted to remaking and refurbishing the developed world's and China's own used electronics. Nobody knows the scale of it, or the revenue, but this I know for sure: there isn't a town, village, or city in China that doesn't have at least one used electronics market. In bigger cities, entire malls, like the one on Shanghai's north side, are devoted to the reuse—rather than the recycling—of electronics.

There are other reasons, too, for Chinese companies to buy what Americans call e-waste (a term that started circulating in the early 2000s), and what the non-English-speaking world tends to call used hardware. The most important is the simple fact that China is home to some of the world's biggest and most prominent manufacturers of electronics, and those manufacturers need gold, copper, and other precious and semiprecious metals, to make new products. One way to obtain those materials is to mine metals; the other is to buy metals recycled from old products that may contain the same alloys as the new ones you're planning to make.

Potentially, there's a lot to recycle. In 2010 China became the world's largest consumer of computers and other electronic gadgetry, and Chinese appliance industry officials now conservatively estimate that Chinese consumers toss out 160 million electronic appliances per year (strictly defined as computers, phones, air conditioners, and washers and driers). That's a breathtaking number, especially judged against the comparatively modest 500 million computers tossed out by Americans between 1997 and 2008, according to the U.S. National Safety Council. Meanwhile, in 2006 the U.S. Geological Survey estimated that Americans were throwing away 130 million cell phones per year—with each ton of phones containing ten ounces of gold. That may not seem like much, but it's far more gold than you'll find in a ton of even high-grade ore. In theory, then, recycling cell phones and other electronic gadgets would appear to be a cheaper, greener means of obtaining gold.

But in practice, that's not so clear. Take, for example, a computer motherboard of the sort that might be found in a laptop, desktop, tablet, or smartphone. At "environmentally sound" U.S. e-waste recyclers, that motherboard will be run through a shredder not so different from an automobile shredder, pulverizing the chips and everything else so they can't be reused or easily refined in home workshops (that's the explicit intent!). From there, the gold-bearing scrap fragments, along with motherboards that aren't shredded, will be shipped overseas to multimillion-dollar high-tech refineries in Europe or Japan—the first U.S. e-waste refinery is slated to open in fall 2013—for chemical and other types of hands-free extraction of the metals. It's a precise, mostly clean method of recycling, but it's also very, very expensive—and it's not done in the United States. Most important, it's incomplete: many of the rarest and most valuable materials in an old smartphone, tablet, or laptop can't be recovered or recycled completely, including several precious metals and the aptly named rare earth compounds embedded in touch screens, vibrating ringers, and other key (if arguably nonessential) features of today's technology. All of that unextracted material ends up being incinerated or landfilled.

There are other ways of accomplishing the same tasks.

In developing countries, circuit boards are often shredded, but before they hit the blades they're stripped—usually by hand—of the precious-metal-bearing chips that cover them. This is accomplished by heating the boards over a hot stove to melt the lead solder that holds the chips to the board. Workers who do this rarely have the benefit of much protection beyond a face mask and goggles. But even if they wear full biohazard suits, the fumes emitted by this process travel far (as of early 2013, many Chinese workshops are beginning to use mechanical means to remove chips, in part to protect workers and the environment). It gets worse: the remaining gold on the circuit boards is removed using highly corrosive acids, often without the benefit of safety equipment for the workers. Once the acids are used up, they're often dumped in rivers and other open bodies of water.

The damage done by low-tech developing world electronics recycling is measurable. A 2010 study in Guiyu, China's biggest and most notorious e-waste recycling zone, revealed that among a cohort of village children under the age of six, 81.8 percent were suffering from lead poisoning. The likely source of the poison was lead dust generated by the breaking of circuit boards and the melting of lead solder. A 2011 study staged in Guiyu showed that 25 percent of newborns had elevated levels of cadmium, a toxic substance that can cause kidney damage, reduction of bone density, and other debilitating effects. Those newborns tended to have parents who worked in e-waste processing industries. Other studies have demonstrated high levels of soil and water contamination, concentrated in parts of Guiyu where e-waste processing is most intense. You don't need to meet the damaged newborns, though, to know something is amiss in Guiyu: every day the government trucks in drinking water to the most contaminated parts.

Guiyu isn't unique. Similar sites exist in India, Pakistan, Romania, Albania, Thailand, Vietnam, and other developing countries. But Guiyu, due to its proximity to raw-material-hungry Chinese factories, is likely the biggest (estimating scale is difficult in an industry where operations are often hidden). Meanwhile, small-scale home workshops proliferate wherever people want gold—including in the United States. On You-Tube, U.S.-produced videos provide precise instructions on how to refine gold from electronic waste using the same "primitive" means used in many developing countries, running up hundreds of thousands of hits. Alas, there's no way to know how many American home e-waste workshops exist (occasional reports of accidents suggest that there are more than a few), but whatever the number, they're an important, essential reminder that so-called primitive methods of recycling aren't just practiced by poor people in developing countries. Somebody at the end of your cul-de-sac might be doing it, too.

That's why I'm at the electronics market. I don't want my perfectly reusable phones recycled using methods that poison toddlers, and I don't want to take them back to the United States, only to have them shredded and sent to refineries in Europe (or suburban garages at the end of cul-de-sacs). In short: I don't want to recycle them. Rather, I want to do the environmentally responsible thing with my old cell phones: I want to sell them for reuse (without having to reuse them myself). I want one of the hundreds of millions of people in China who make less than $5,000 per year to have access to my old devices.

Reduce. Reuse. Recycle. Like most Americans, I don't like the first verb, so I do my best to practice the second one. It's better than the third, especially when the unwanted objects are electronics.

I move along to another booth, where two spiky-haired young men labor together over an open laptop with a voltage meter, checking connections. One holds a CD-ROM drive in his hand which, I think, is about to go into one of those computers. He looks mostly free, so I stop and offer him my clear plastic bag full of cell phones. He looks up, his eyes narrow at the motley sack, and he shakes his head.

Fine.

I move along to the next booth, where a tough-faced young woman, maybe twenty-five, takes a quick glance at my bag and shakes her head. I linger for a moment, and I guess I see her point: she has phones in her display case, yes, but none look nearly as old as mine. "Three-G," she says with a nod. "Three-G."

My phones, for sure, lack 3G capability. They're just good old solid dumb phones. If a migrant construction worker needs the ability to cruise the Internet with a phone, then that migrant should look elsewhere. But if he wants to call his mom on the weekend, then my bag contains his solution. The problem is, it seems like everyone at the market—even the students and migrants who frequent this market—want 3G. Inside the display cases, all of the phones have Internet capability. Old dumb phones like mine are turning out to be as relevant, as marketable, as a VCR in Manhattan.

I walk out the door and wander down the street, past several shops selling used computers and other refurbished goods, until I come across two cardboard cartons stuffed with cracked plastic computer cases. A few steps away I find another box, this one filled with old computer motherboards. I consider dropping my phones into the box and having done with this obviously pointless quest.

Then I reconsider. Those boards are almost certainly bound for Guiyu and its toxic workshops. More likely than not, the owner of the box wouldn't mind if I dropped my phones into it so they can hitch a free ride down there.

But then something else occurs to me: I'm a scrap journalist! So, rather than let some random circuit-board salesman dispose of my phones for me, why don't I travel to Guiyu and sell them myself? It's a preposterous idea—white people don't just show up in Guiyu hoping to sell a few cell phones—but as I ride the subway home, I decide that I'm the man for the job. All I need is somebody to take me there.

When China began importing foreign scrap metal in the early 1980s, used electronics were packed into many of the containers. For example, one U.S.-based exporter told me that he made a fortune by exporting container after container of old analog telephone equipment in the early to mid-1980s as the telecommunications industry upgraded to digital equipment. Other scrap dealers, including Tung Tai's Joe Chen, have told me that they began exporting old mainframes to China in the mid-1980s, and IBM and Apple PCs as early as 1985. The reasons for doing so were simple: the United States didn't have an electronics scrap recycling industry in those days. But even if the United States had been home to businesses capable of profitably breaking an old telephone into its various recyclable pieces—plastic, copper, steel—it still lacked something important: manufacturers interested in using all of that old plastic. As a result, American landfills are stuffed with electronics from the 1960s, '70s, and '80s.

Enter Taiwan, and then China: labor was inexpensive and, more important, growing economies demanded everything that could be extracted from an old telephone or mainframe. So that's where U.S. recyclers sent the old telephones that would go to the landfill: Asia, where not only did workers separate plastics from metals but there were companies eager to use the plastics (though not always in ways that would meet developed world environmental or quality standards).

Sending electronics to Asia is often characterized as "dumping." It's a powerful phrase designed to evoke an image of trash dropping down from the wealthy heights of the mostly developed world to the impoverished depths of the developing world. But there's one major problem with the phrase: it implies that by dumping in Asia, a shipper is saving by not recycling properly at home. Nothing could be further from the truth. Even in 2013, it's still legal to dump many types of electronics into American landfills for no more than it costs to toss a bag full of Burger King wrappers of the same weight. So if it's all but free to dump electronics into landfills, why on earth would anyone turn around and *pay* to ship those same electronics to China?

The answer is simple: the value of those electronics exceeds the cost of shipping, and people in China are adept at extracting that value. However, the value isn't scrap-metal value. In most cases, a computer monitor has no more than $2 or $3 worth of scrap packed into it—barely enough to cover the cost of shipping the nine hundred monitors you can pack into a container from Minnesota to Shenzhen. Rather, the value—and the business potential—is in reusing whole machines, or parts of them.

Even back in the 1980s, old, so-called end-of-life computers exported from the United States had quite a bit of life left in them—especially if you looked at them from the perspective of someone whose only hope for mechanical or electronic computing help was a slide rule. Think about it: if you are poor and Chinese and didn't have access to new computers or calculators, a five-year-old IBM PC is still better than no IBM PC. And if in 1990 you are a poor Chinese scientist at a university, a gently used PC is a borderline miracle. Some old computers could be reused (and sold) right out of the shipping container; others were broken and had to be repaired, sometimes using new parts, sometimes using old.

That's not dumping.

Rather, from a Chinese perspective, it's an opportunity to take advantage of a price difference between two markets. On Wall Street, they call that arbitrage. At the neighborhood garage sales I used to frequent with my grandmother, it's called a steal, akin to buying a $200 antique for a quarter. And in Shanghai, they call it the source for much of the inventory at the used electronics markets.

It was (and is) a good business that enriched businesspeople and the coffers of local governments alike, but by the early 1990s Guangzhou and Shenzhen—where it thrived—were starting to emerge as major metropolises. Local governments were justifiably wary of the pollution associated with e-waste processing, and encouraged a mass relocation. So China's e-waste traders and processors started looking for somewhere remote and likely to remain that way. The place they found was a set of mostly inaccessible farming villages located in the mountains of northeast Guangdong Province named Guiyu. The shift took place quickly: by the early 1990s, Guiyu become the biggest e-waste processing hub in southern China.

Guiyu has never been a secret, either in China or abroad. Hong Kong media covered its horrors in the early 2000s, and Communist Party–controlled Chinese media has devoted considerable coverage to it since at least the mid-2000s. Outside of China, Guiyu entered the popular consciousness in 2002, after Jim Puckett, a Seattle-based environmental activist, visited the town and published his findings in *Exporting Harm: The High-Tech Trashing of Asia*. The report included graphic photos and descriptions of low-cost, unsafe electronics recycling. For example,

Puckett described the acid stripping of copper from circuit boards in these terms:

> This mixture and process was invariably applied directly on the banks of rivers and waterways . . .
>
> The process resulted in huge clouds of steamy acid gases being emitted, which looked like smoke from even far away. Worse, the process resulted in the routine dumping of *aqua regia* process sludges that blackened the river banks with the resinous material making up computer chips . . .
>
> The men worked at this process day and night protected only by rubber boots and gloves. They had nothing to protect them from inhaling and enduring the acrid and often toxic fumes.

Puckett's account was an instant media sensation that raised the previously unknown problems associated with the disposal of old electronics into a top environmental priority (in the United States and Europe, at least). Nonetheless, more than a decade later Guiyu not only persists, it thrives openly, mostly unmolested by local and national Chinese governments. How open? Stop by the Chinese-language website for the Guiyu Resource Recycling Association (www.guiyu.org), a Communist Party–approved trade group for Guiyu's recyclers, and on the home page click "A Precise Description of the Guiyu Dismantling Industry." The first paragraph reads:

> (1) the source of the waste electronics (e-waste):
> Waste electronics (e-waste) come from Japan, the United States and other countries through Hong Kong and Taiwan, and then enter through Shenzhen, the South China Sea (ports), Guangzhou and other places. Guiyu buyers (sometimes through an agent or principal) will bargain for the goods and then transport a container to their own Guiyu factories and workshops where they are disassembled.

How can that be? Doesn't the leadership of the Chinese Communist Party in Beijing care? Don't the good-natured regulators (I've met many) at the Ministry of Environmental Protection pay attention?

I ask the question of an Asian-American scrap-metal processor who—over dinner one night in 2011—agrees to take me to Guiyu to sell my phones. For the purposes of this story, I'll call him Henry. He's a significant exporter of scrap—including electronic scrap—from developed-world countries to China, and he maintains tight connections to China's environmental policy makers, regulators, and customs officials, as well as state-owned companies that process electronic scrap and are keenly interested in building developed-world recycling facilities across China. "Domestic [that is, Chinese-generated] electronic scrap is getting bigger," he tells me. "And the central government wants it to go to one place. They want a designated zone. They don't want this kind of polluting business popping up all over China. They want it in one place."

In other words: Guiyu is what China's central government has designated as *the* place where they'd like to see China's electronic scrap recycled.

This isn't just the idle speculation of Henry, a self-interested scrap dealer. In 2004, representatives of China's National Development and Reform Commission (NDRC) arrived in Guiyu to evaluate the environmental disaster for themselves (helpfully, the Guiyu Resource Recycling Association has posted pictures and accounts of their visits to its website). This was significant: the NDRC is one of China's most powerful government agencies, charged, since the early 1980s, with reforming and restructuring China's economy into a modern, market-oriented success. It's impossible to say what they thought of what they saw (NDRC officials did not respond to my request for an interview), but this we do know: in 2005 they, in concert with six additional high-level Chinese government agencies, announced that Guiyu would receive significant funding to upgrade its facilities in line with China's new emphasis on sustainable development. According to the government's announcement, officials would "accelerate the construction of Guiyu into a national demonstration base for recycling."

There was one problem: the Chinese hadn't started tossing out their own computers in great enough numbers to sustain Guiyu's recycling industry. So even though Chinese customs regulations prohibit the imports of used electronics, somebody somewhere made sure that Guiyu continued to receive imported electronics. This is a point made by one of Henry's partners the night before we depart for Guiyu. According to him, China now generates more than half of all of the electronic scrap that turns up in Guiyu, and the percentage is growing—fast.

And there are other reasons, still, that the toxic trade persists.

In 2007 a young researcher connected to China's Ministry of Environmental Protection, who has spent years working on China-appropriate solutions to the e-waste issue (in the absence of technology shares from American, European, and Japanese recyclers), offered another reason Guiyu persists. "In China, most people don't even have safe food to eat, clean water to drink, and clean air to breathe," he told me. "We are perhaps ten or twenty years from solving those problems. But the foreign environmental groups want us to worry about old computers and greenhouse gases. How can you worry about greenhouse gases and old computers if your kids don't have safe milk to drink?"

It's an important point that has wider implications. In the United States, the European Union, and other developed regions, the environment is clean enough that citizens and governments can spend time and resources worrying about how to recycle old iPods. But in China the environmental and health challenges are so great that cleaning up Guiyu won't make any appreciable impact on the environmental condition of the country as a whole. Personally, as a ten-year resident of Shanghai, I'm far more concerned with the abysmal air quality and regular food-safety scandals than I am about what happens to my old phones in a town in southern China—and I'm professionally connected to the last issue. In any event, even if China actually followed through on its halfhearted commitment to banning the import of old electronics, Guiyu would persist on the quickly growing tide of electronic devices being tossed out by the several hundred million Chinese who are already middle-class. According to Professor Li Jinhui of Tsinghua University in Beijing, China generated 3.5 million tons of e-waste in 2012. By comparison, in 2011 the United States generated 4.1 million tons of e-waste, according to a comprehensive 2013 study by the U.S. International Trade Commission (and more than 80 percent remained in the United States!).

Is that a lot of e-waste? That's open to debate.

In 2012 e-waste amounted to less than 1.4 percent of all waste generated in American homes and offices. In the grand scheme of American wastefulness, that's not very significant. But allow me to place it in further context: in 2012 the United States generated more than 36.43 million tons of food waste, according to the same EPA study, or ten times more food waste than e-waste, by weight, that year.

So why, then, does e-waste command so much attention when we talk about recycling? Why do news organizations like the BBC regularly refer to the so-called e-waste crisis, when there is clearly a far bigger American food-waste crisis that receives almost no attention?

Activists and others will argue—correctly—that e-waste contains hazardous substances that need to be handled properly when recycled. When they aren't handled properly, they can cause environmental and safety problems. I agree—but only to a point. After all, plenty of other recyclable products have similar problems when it comes time to recycle—paper to plastics to copper—and yet they're rarely the focus of activists. We don't have a "waste-paper crisis" (and China has been struggling for two decades to clean up polluting paper mills) or a "plastic-bottle crisis." Instead, we have an "e-waste crisis." Why?

In my opinion the answer is simple: any industry connected to IT in the twenty-first century will automatically receive an outsize portion of the media's attention (scrap iPhones are sexy to reporters; rotting, unfinished piles of food are not). It helps, too, that the global IT industry has been among the strongest supporters—financially and otherwise—of efforts to prevent old and used electronics from being shipped to the developing world (Samsung and LG have been particularly active in this arena). Thus what was once a perfectly repairable old computer (which competes against the purchase of new computers) has over the last decade been rebranded as "hazardous e-waste." That's not good for consumers in the developing world, or for the environment.

From the perspective of someone who has visited recycling facilities across the world, I think it's worth recalling that developing countries often have priorities that differ from those of activists in Europe and the United States. It's a point that was pounded into me in March 2013, during a visit to Toxics Link, a nonprofit environmental organization in Delhi, India, that has expended considerable efforts trying to educate the city's thousands of small-scale electronics recyclers about the dangers inherent in using toxic chemicals. Priti Mahesh, a senior program coordinator with Toxics Link, told me that their education efforts hadn't been successful, and she didn't expect that they ever would be. "[The e-waste processors] will tell you, 'I can't worry about health effects twenty years from now,'" she explained. "'If I don't do this work, I'll starve and die tomorrow.'" Later in the interview she concluded: "You can't always be so idealistic."

I heard something similar a few days later in Bangalore, India's Silicon

City, home to one of the world's most vibrant IT industries, and a thriving, informal, unsafe, and polluting e-waste reuse, repair, and recycling industry. There I sat down with Dr. Vaman Acharya, chairman of the Pollution Control Board in the state of Karnataka (home to Bangalore), to discuss e-waste. Thinking about the scale of poverty evident across this growing city, I asked him whether e-waste was his most pressing issue.

He smiled. "No!"

"Then what is?"

"Garbage," he answered, and went on to explain that he's struggling to figure out how to handle all the nonrecyclable waste—mostly food waste—that households generate in his jurisdiction. His second concern was sewage, followed by industrial pollution, dust particulates generated by construction, biological wastes from medical facilities, plastics disposal and recycling, and the management of chemical wastes. E-waste was listed only at the end, after he couldn't remember any other priorities. "I have much bigger issues!" he exclaimed.

It was a clarifying lesson in what matters to the developing world.

One afternoon in late 2011, I join my friend Henry and four Guiyu recyclers for a late lunch in a restaurant below a highway overpass in Humen, a textile manufacturing hub roughly forty-five minutes into the five-hour drive from Shenzhen to Guiyu. I'm here as the unannounced guest of Henry and his Foshan-based partner, whom I'll call Du. It's a touchy situation: the Guiyu Resource Recycling Association may take pride in hosting high-level Chinese officials (they post photos of them to their website), but the town's traders aren't keen to have yet another foreign journalist show up to film the worst of what the town has to offer. Such visits invariably result in show-trial shutdowns of local businesses.

Henry doesn't waste any time introducing me to the Guiyu gentlemen. According to him, I'm the son of an American scrapyard owner, and I'm interested in learning more about Guiyu's electronic scrap industry. The proof is in my backpack: in addition to five old cell phones, I have an old HP PDA that once belonged to a friend in Minnesota, a 2008 Dell laptop that failed on me just over a year after I bought it, a tangle of old cords, and an old LG phone charger. The hardware isn't my only reason for being here (I have a book contract, after all), but it's enough truth to be very welcome news to the Guiyu boys. Americans have long been among the leading exporters of electronic scrap to Guiyu.

The traders look back and forth at each other. One shrugs, takes a drag off his cigarette, and clears his throat. He's a freakishly thin man with a scarred eyelid that I'd like to think he acquired in an e-scrap processing accident. "It started probably twenty years ago. The e-scrap was imported into Nanhai. Some people from Guiyu were down there and brought it to Guiyu, and made money. Others saw him make money and joined the business. That's how it grew." I look around the table, but nobody disagrees or has anything to add. Instead, they ask Henry for more details on the business he's interested in doing on this, his most recent visit.

Henry has an interesting proposition for them: he wants to buy computer chips *before* they're refined for gold in Guiyu's notorious toxic workshops. He has a client, he says, in Japan, who wants to refine the chips there. The traders all nod: Guiyu specializes in extracting gold from computer chips, but they lack the technology to effectively extract the other precious metals, including platinum and palladium. Because of this, one of the Guiyu traders tells us, the Japanese have been showing up to buy gold-bearing chips in Guiyu for at least a decade, now. Guiyu's traders, knowing that home workshops can't hope to compete with Japanese technology, have happily taken on the role of dismantler and supplier of chips to Japan. This doesn't sound like dumping to me; it sounds like a very real, very well-established supply chain. It's also not unusual. India's electronics recyclers export vast amounts of precious metal bearing electronic scrap to refiners in Belgium for the same reason: high-tech refiners can make more from—and thus will pay more money for—computer chips extracted in developing world workshops.

Lunch is brief, and soon we're back on the highway, headed north to Guiyu. We're accompanied by a young processor and trader, who I'll call Ge. He's no older than his late twenties, boyish and likable. His is not the presence of someone who engages in what some have characterized as a toxic trade. Rather, he's soft-spoken, with a young man's sincerity of purpose. He has no idea I'm a journalist, and I suddenly feel a twinge of guilt for deceiving him.

The drive is uneventful, and for much of it we follow a highway that hugs the coastline. We pass fishing villages, boats, and buoys marking the location of nets; gas stations, repair shops, and truck stops that house McDonald's. Halfway through the drive Henry wakes from a nap, takes a look at my old Samsung phone, and tells me: "Okay, see this phone. Maybe it's a 1999 phone. It's a Samsung." His eyes light up. "So for you maybe it's scrap. But I know because it's a 1999, I know that it has a certain kind of chip that I can sell for a certain price. Maybe I know that the screen has a different value. And maybe I know there's memory in it, too, that has a value. So I can see more value in it than you can."

"Who buys it?"

He laughs. "Somebody who wants to use the chip again! Many companies that make scrolling digital signs, they like these older chips. They can run that application for a long time."

In other words, the chip in my old Samsung might be extracted and then transplanted into a scrolling digital sign purchased by a Kansas diner to advertise the daily lunch specials. It's a downgrade from running spreadsheets, Web browsers, and games, but it sure beats digging up gold, copper, and silicon for a chip.

"How long can a reusable chip run a sign?" I ask.

"Hard to say," Henry answers. "Maybe fifteen years."

That's better than shredding and recycling, it seems to me.

For Guiyu's traders, according to Henry, it's the reuse value, and not the scrap metal and plastic value, that really drive profitability. Think of it this way: Guiyu's processors buy old phones by the ton, and prices might range as low as a penny or two for a device that contains perhaps a few pennies' worth of gold, copper, and plastic. But if, as Henry implies, an old phone contains a chip that can be resold to a manufacturer of signs for $10? That's a profitable business. "The reuse is maybe eighty percent of the profits in Guiyu," Henry tells me. "Huge, huge, huge business."

This isn't a total surprise. In 2009 I visited EconEcol, a Japanese scrap recycler with several warehouses at the base of Mount Fuji, one of which is devoted to dismantling old pachinko machines (the Japanese equivalent of a slot machine) purchased from some of Japan's thousands of gambling houses. As I walked the warehouse with one of the company's managers, he told me that gaming machines' diminutive HD touch screens are carefully extracted from the other components, packed, and then shipped to China, where they're installed in GPS units. Pachinko screens, apparently, are just the right size for fitting onto a dashboard. What's more sustainable than that?

It's late afternoon when we arrive in Puning, a dense city of 1.5 million that abuts Guiyu. I know the name well: a few years ago the town had well-publicized plans to sterilize 900 women to meet population control goals. Then the central government in Beijing, bowing to public outrage, intervened. Governments around here, it turns out, do as they please until they can't.

As I stare out the window, we pass row after row of densely packed, crumbling concrete apartment buildings. Traffic is as dense as a Shanghai rush hour on the main boulevard through town, but far more dangerous (which is saying something): there are no traffic lights, so kids on bikes wait for gaps in traffic to shoot across the road.

This is a murky place, lit by kilometer after kilometer of street-level commerce, ranging over restaurants, convenience stores, hardware stores, and the occasional streetlamp. Every few minutes the new sprawl ceases and gives way to the gently arched roofs of ancient village clusters. They're shadows in the night, unlit and abandoned, the recent remains of the farming life that was once this latest boomtown's mainstay, the old China's modest meaning. Nobody has had time to demolish them yet, I guess. Or maybe nobody has the heart.

Those abandoned villages are better hidden than Guiyu. In fact, the notorious electronics dumping zone isn't hidden at all; it's not remote, difficult to access, or even hard to find. Rather, it adjoins this sprawling new metropolis, reached—after thirty minutes of traffic—by driving over a short, arched bridge that rises over a dark canal. "Once we get to the other side," Henry tells me, "you've made it."

A bright flash suddenly illuminates the van.

"Our picture," Henry adds. "Every car entering the town, they photograph and keep it for a month."

Maybe the government around here isn't as relaxed about outside attention as the Guiyu Resource Recycling Association's website seems to suggest.

It's night, completely countryside-black (thus the giant flash spy camera). I see a few three-story buildings glowing in the darkness, and then we turn hard right into an alley and stop. To our left is an unlit sign covered in Chinese characters and two members of the alphabet: IC. That stands for "integrated circuits," the chips that make the world's devices hum, and make this notorious outpost in the global recycling trade very, very rich. New ICs from Intel, Samsung, and other manufacturers are worth hundreds and sometimes thousands of dollars; by the time they reach Guiyu, however, they're priced by the kilogram, and rarely worth more than 30 cents apiece.

I follow Henry, Ge, and Du out of the van and into a clear, starry night. But there's something unnatural out here, a thin, murky smell—chemical, like melted plastic, and vaguely sweet, like chrysanthemums. As it fills my lungs, I shorten my breaths; thankfully, we're only planning to spend a night and a day.

Across from me, across the road, are a few large plastic sacks filled with rectangular PC cases—and then . . . distance. Farther out, I see the glow of what I presume is Puning. "A few years ago," Henry whispers to me, "One mu [.16 acres] over there was worth eighty thousand dollars. Now, one million."

It's no mystery why. In Guiyu, the recyclers have so much money and so few places to spend it. So they buy real estate.

Henry nods at a high, spiked gate that blocks the entrance to a dimly lit warehouse. "This guy's son owns that land," he says, pointing across the road to the empty space.

"This guy" turns out to be a bony man in his early sixties. As he approaches the gate in the shadows, his face is pixieish, with an afterthought of a goatee and mustache. Du, Henry whispers to me, has been doing business with the old man for years, now, sending scrap from Foshan to Guiyu. From Guiyu, it goes back to Shenzhen and is remade into new electronics. It's the common path trod by scrap metal from the developed world to China. I don't have to think twice about it. Meanwhile, my eyes are drawn to the giant white teeth that shine behind the old man's warm smile of recognition. They must be dentures, I think, purchased with the proceeds of IC.

I watch as he uses a key to unlock the gate from the inside. Once we're inside, he relocks it and leads us into this room—his lair—his bony knees brushing against each other, his bony elbows swinging.

The security, I quickly realize, makes sense: spread out before us are thousands of boxes in haphazard piles that fill a space the size of a hockey rink. It's all dimly lit by a handful of fluorescent bulbs that quiver in the night. I see motherboards, pieces of mainframe comput- ers, and hard drives; electrocardiogram units, keyboards, laptop fans, and screens. But what strikes me immediately is that almost none of the so-called e-waste looks used. Instead, it's wrapped like new.

As I look around, I don't find anything that looks like a computer that somebody might have once used in their office, a laptop that once belonged to a student. For example, those HP laptop screens aren't old and dirty; they're still wrapped in cartons labeled "HP." Those rolls of Samsung computer chips aren't burned and lying in trays; they're in new Samsung boxes. That box of Panasonic screens in the corner? They're wrapped in plastic, with individual pink slips labeled "unserviceable" and "Panasonic Avionics Corporation." Did Panasonic send them? Did HP sell them? Or did they arrive via other means? Is this what Henry meant when he told me about the reuse market? Are these screens reus- able, too?

There's no way to tell without asking, and for now I'm keeping my mouth shut. But the fact that they're here, and not somewhere else, makes abundant sense. After all, the manufacturers named on those boxes are located four or five hours south of here, in Shenzhen, in Dongguan, in Zhongshan—the cities where so much of the world's electronic gadgetry is manufactured. The manufacturers of this stuff are also located in Malaysia, Taiwan, Singapore, and Thailand—places from which trans- portation to Guiyu is quick, cheap, and easy. Their surplus is Guiyu's inventory to be processed; their defective boards, monitors, and chips are Guiyu's opportunity for huge profit margins.

Consider, for example, that a box of defective or surplus laptop mon- itors is more than just a box of monitors—it's a box of parts, only some of which don't work. If you have dozens of those boxes, then likely you have hundreds of identical parts that can be extracted and resold. If you can extract and salvage two hundred of a certain microprocessor, then you have the ability to sell a bulk order of that microprocessor to a remote-controlled toy car manufacturer up the road. Thus, a processor purchased for next to nothing becomes a part of a product worth per- haps $100.

Alas, there isn't much time to look around. The old man is gesturing for us to join him on a wooden sofa that lacks cushions. He takes a seat and tucks his bony legs up to his bony chin. It's sinister here in a noir kind of way, but maybe that's just me: Du and the man are talking of their families, of Chinese New Year, of a favored Hunanese restaurant in Shenzhen. The old man pours a freshly brewed pot of tea into tiny cups that he hands to each of us using a steel claw. I take mine and sip it: the sweetness glides across my tongue, warm and complex, rare and unquestionably expensive.

I look around the room. Next to us, against the wall, are monitors that broadcast black-and-white images from cameras arranged in front of the warehouse and gate. They cast a grainy light that the old man watches all night long.

"Five years ago," Henry whispers in English, "he was just a small guy. A little farmer. Now he's huge. His son operates a shop in Shenzhen where they sell the chips they get from here. Good business." It occurs to me that this old man is responsible for the actual reuse or recycling of more computers in a week than the average San Francisco neighborhood manages in a decade.

Nonetheless, the old man is not oblivious to the problems associated with his trade. With that big, toothy smile, he tells us that a high-level delegation from the central government in Beijing recently visited town. According to him, they've already provided $80 million to clean things up and move Guiyu's dirtiest processing off the riverbanks and into indoor workshops. To continue financing that alleged upgrade, the town of Guiyu is collecting a fee of approximately $11,000 from each of its roughly 5,500 electronics recycling workshops.

But if the additional $60 million raised by that tax is not enough to clean up Guiyu's polluting ways (and I don't think it comes close), the government allegedly has a wealthy and powerful new partner to help things along: TCL, one of China's largest consumer appliance manufacturers, and a company with deep roots in the People's Liberation Army. Details on TCL's investment and role are sketchy, but according to plans that I was shown, the company is playing a key role in building an industrial park where recyclers will be forced to locate. There, they'll be required to use upgraded technologies for chip extraction, refining, and other types of recycling. In return, TCL—a company that uses metals and plastics—will have the right of first refusal on certain raw materials generated in the park. When I contacted TCL about the investment, however, they declined to comment.

As for the old man, he doesn't care either way. He and his son have a successful business that's enriched the family well beyond their agricultural roots. If and when the local and national governments make the decision to cut out the small entrepreneurs who built Guiyu into a global electronics recycling hub, he'll be disappointed, but it won't be a total disaster. The government and its partners might have a monopoly on the recycling of electronics, but they'll still need men like the old farmer to source the electronics from around China and the world, and to sell the reusable components once they've been extracted.

For the old farmer, it's the reuse of chips that makes the real money. According to Henry, two other sources, and the Guiyu Resource Recycling Association, the top customer is in fact Chenghai, a nearby town nicknamed Toy City due to its high concentration of toy manufacturers. Many of the toys manufactured there are electronic, and they require microprocessors of the sort that are recovered and sold in Guiyu. According to the trade association, Chenghai is customer number 1 for the products of Guiyu's toxic trade. Think about that: somewhere, a parent is giving a child a toy made from used computer chips extracted in one of Guiyu's notorious workshops.

And that's why Guiyu will survive: China's economy is too dependent on what Guiyu makes. The old man's real estate investments in-town are a sure bet that the town's favored industry will only grow.

As we prepare to leave for our hotel, the old man rushes into a back storage room and emerges with a carton packed with coffee-cup-size boxes of tea worth, according to Henry, $80 each, and hands it to Du. Henry whispers that the box probably ran $1,500, but it's no matter for this rich old man. "For Chinese New Year he bought an entire truckload of fireworks. Blew them up for days. That old farmer, he's rich."

The morning of my first full day in Guiyu begins in the van with Henry, Du, Ge, and two of Ge's cousins. We drive through a drought-stricken landscape beneath a clear blue sky of the sort that I don't often see in polluted Shanghai. Puning is only a few miles away, but its cranky bustle couldn't be further from these quiet fields and country roads. That might soon change, though: Ge slows to show us the hulking concrete pylons and tracks being laid for China's national high-speed railway system just outside town. It's a key program of the central government in Beijing, designed in part to boost the fortunes of China's countryside. "There will be a station two miles from here," Ge says. "Good for business in the area."

It's difficult to say just how good for business. Over the years, journalists and environmental organizations have made estimates and guesstimates of just how much electronic scrap is processed in Guiyu. Yet the fact of the matter is that the only people who have a good fix on the scale of the business belong to the local government that benefits so richly from taxing it. In the absence of a disclosure from them, the best, most credible (and arguably most self-serving) data available is that published by the Guiyu Resource Recycling Association. According to them, the twenty-one villages that constitute Guiyu are home to more than 300 private companies and 5,500 family workshops that employ more than 60,000 people. On an annual basis, they dismantle and process 1.55 million tons of waste electronics. From that, the industry extracts 138,000 tons of plastic, 247,000 tons of iron, copper, aluminum, and other metals, and a jaw-dropping 6.7 tons of precious metal.

As we drive, I glance at a large cluster of seven- and eight-story textile factories along a river. Ge says that fashion and textiles are a major industry in Guiyu. I don't recognize any of the brand names on the signs, but I do recognize the products: socks, bras, shirts, trousers. It's funny; last night we ran into some Italian designers who were in Guiyu to work with their contractor on some new lines. We chatted briefly, and in the course of the conversation they asked whether I was there to supervise production of my company's "spring lines." They didn't seem to know that Guiyu's most famous business isn't the one that they do, but rather the one that has transformed parts of Guiyu into a toxic wasteland. How could they miss it? I guess the better question would be: Why would their contractors ever tell them that such a business operates here?

Guiyu, the town, is actually a recent invention, the result of rapid economic growth that's connected once-separate centuries-old farming villages via narrow streets, grimy high-rises, and tight, raucous street-level commerce. I don't see any of the black or orange smoke described by the activists and reporters who made Guiyu famous. Part of the reason is that much of that smoke was produced by burning insulation off wire, but the rise in oil prices, and the development of markets for used insulation, has definitively ended that business.

But there's another reason, too. According to Henry and several

Guiyu traders, one of the first-stage goals is to move the industry indoors, where it's more difficult for outsiders to see. Those outsiders include activists, journalists, and—increasingly—the residents of rapidly expanding neighboring Puning.

As we drive through Guiyu's downtown, I'm most startled by how normal it seems. To me, it's just another small Chinese city of four- and five-story buildings with street-level commerce. But soon I notice a difference. First, an unusual number of storefronts advertise "IC"—integrated circuits like Intel Pentium processors—for sale. And second, and more ominously, every third building or so has a small tool shed attached to the front of the building, next to the door. Depending on the size of the shed, one or two round metal smokestacks rise out of the walls and shoot up to the roof. The stacks remind me of cobras ready to strike, and every other building, it seems, has one.

Soon I realize that there are hundreds of these stacks in downtown Guiyu—maybe more—and here and there they trickle smoke. But not much: Ge tells us that there are "simple water filters" on them that keep visible pollution to a minimum. "We get in trouble with the government if we burn too much," he adds. Henry adds another detail: "They probably burn at night." That would explain, I suppose, why the air smelled so strong outside the old man's warehouse in last evening's darkness.

When we finally step out of the van, the sweet stench is present, but not nearly what it was.

We've arrived at the workshop of a wealthy computer chip trader. As we walk into his workshop, I'm surprised to see young children riding their bicycles on the concrete next to a workbench where thousands of small circuits are piled on a table, waiting to be sorted by type into fifty small red bowls. To the left are shelves and display cases holding hundreds of bags—and tens of thousands of computer chips—for sale to whomever stops by.

But what Henry wants me to see is beyond a metal door protected with two deadbolt locks in the corner. It's open, and he leads me inside and suggests that I take a photo, quick. On the floor is a pile of circuit boards stripped of their chips, and smaller piles of chips stripped from circuit boards. Across from them is heavy-duty industrial oven and stovetop, roughly the size of a large microwave. It's covered in melted solder. A pair of pliers, a pair of pincers, and a box cutter sit atop an old paving stone next to it. Another stone is covered with the bright silver streaks of more melted solder. I look up, and there's the round hole that leads to the filter, and the sky. It's medieval and vaguely modern, computer chips and pincers, and it feels ominous, dark.

"Hey, hey, hey!" The owner of the place has just walked down a set of stairs in the main room. He's a short and squat man—a farmer's body—with a particularly mean-looking baby face. He could be thirty; he could be fifty; I don't know. His kids and wife circle around him, wondering who the foreigner is, and he waves them away. Fortunately, Ge steps forward to explain everything. As he does, Henry glances at the burn shed and shakes his head. "Would you raise your kids here?" He whispers. "My god."

I take a deep breath. My god.

But there's no time for that kind of talk now. The mean-looking baby face is signaling for us to have a seat around a coffee table where he pours tea. He only glances at me, and the glance is cold: he doesn't want my white, foreign face anywhere near his business. I can imagine that his bullish face would be in mine if I were alone; I can imagine violence. But I'm here with Henry, and Henry is technically backed with big money from Japanese refiners. The baby face yells for his wife to pull some bags off the shelves. She does as told and places them on the coffee table.

The baby face lights a cigarette and then opens the bags. One by one, in a cold, low voice, he tells us prices. Intel's Pentium III chip goes for around 30 cents per chip (I remember when they went for hundreds of dollars, new, and I coveted a PC built around one). The gold content, Henry explains, is high. So it goes: chip after chip is priced out, mostly by the kilo. Henry writes down the prices in a notebook, stashes it in his briefcase, and we leave.

From there we drive ten minutes to the Guiyu Electronic Components Market—a six-story wedding cake of a building. It's nothing like Taizhou's reuse market, though. This one is on the edge of town, set amid the open plots of abandoned farmland intended for Guiyu's mandated government cleanup and transformation into a modern national recycling hub. Henry points at some long buildings across the street. "Those are the big operators moving into Guiyu," he tells me. "The small guys are going away. The big guys will go into those buildings."

"Can we visit?"

He smiles. "Nah."

The market's parking lot is full of motorcycles, each with a small carrying box attached to the back end that—I imagine—is just big enough for a bag of computer chips extracted in a burn shed overnight. I expect something wild, like the bustle of the vegetable markets near my Shanghai apartment, but instead I find quiet rows of display cases filled with chips and other high-tech components. We walk the aisles, occasionally stopping to admire a bundle of old Intel chips, a bundle of old Motorola chips, a stack of circuit boards as clean as new.

You can buy one Intel Pentium III chip here; you can buy hundreds in bulk. And those bulk orders of chips don't come from home PCs "recycled" in the United States. Rather, they come from containers of obsolete computers sent in bulk to Guiyu from businesses in China and all over the world; they come from defective motherboards that a manufacturer sold in the process of cleaning out a warehouse.

None of these chips are going to Japan for refining. Rather, they're all destined for reuse in new products. That's why it's so quiet in the late morning: factories buy chips early in the morning and late in the afternoon.

The market is such a clean and orderly place that it's hard to imagine anyone being poisoned by the extraction of the chips that are sold here. But there are hints of darkness. I notice vendors twitch and reach for their iPhones as we pass. When I take pictures of what's to be found in the display cases, they make furtive, soft-spoken phone calls. Later, when I take a picture pointed down the aisles, vendors jump out of the way of my lens. And finally, after ten minutes of browsing, a skinny young man in a black T-shirt approaches Ge and begins whispering something to him. Ge nods and walks over to Henry, who rolls his eyes. "I wanted to take you to see the mayor," he tells me. "But I don't think we can do that. Better get going."

I watch over Henry's shoulder as the skinny man opens a vintage Motorola Razr phone and makes a phone call. His head is bent downward, and he looks like he's reporting something to somebody. I feel a chill; I don't belong here. It's one thing, after all, to offend the mayor of Guiyu; it's another altogether to offend whomever the National Development and Reform Commission stations to protect this place.

Outside, Henry lingers to chat with Ge and two vendors who clearly want to do business, regardless of my presence. But even if they didn't, I suspect Henry would take his time. He's not the sort to be intimidated, even in Guiyu. His connections are too good.

As he deals, I walk into the landscaped square in front of the reuse market. There, at its heart, I'm astounded to find a rowboat-size statue of a gold ingot. It's not a brick, however, but an ancient Chinese sycee, shaped like a bowl with a scoop of ice cream in the middle. The sycee shape hasn't been used as money in more than a century, but it remains a potent symbol of good luck and prosperity across China. During holidays, parents buy sycee-shaped chocolates for their children; in taxis, drivers hang plastic sycees from their rearview mirrors.

So what does it mean to place a giant sycee in the middle of a city square? I'm really not sure, but the red pedestal upon which it sits offers a hint. On it, gold paint spells out a single character: 聚. *Jù*. It means: assemble, gather, join together. In other words: gather here, and pay homage to gold.

Guiyu's traders recover an astonishing five tons of gold every year from the electronics that are transported into town. *Five tons.* That's the weight of a large delivery truck dribbled out in quantities the size of a raindrop, from computer chips bathed in acid that turns the eyes of old farmers into glassy pairs of yellow marbles.

Henry walks up beside me. "Let's go." But the golden ingot gives him pause. "Jesus," he says. "Look at that."

It's early afternoon when I ask Henry if I'll still have the chance to sell my e-waste. So far, I don't feel like I've encountered anyone who would care to buy it. He assures me that I'll have that chance when we drop in on Ge's family.

Ge's home is on a narrow dusty lane and protected by a high concrete wall and a heavy steel gate that requires several keys to open. We step past it into a small courtyard filled with stacks of old desktop PCs, monitors, burned circuit boards, and a fish cage full of dismantled cell phones waiting to be divided into their parts. Off to the side is a burn shed. Before I can get much of a look, I'm directed into the house. At the door, I remove my shoes in favor of slippers, and climb the stairs to a spacious family room.

Ge points at an orange pleather sofa, and I have a seat. His mother emerges from the kitchen with a plate of freshly cut watermelon and giggles at me as she places it on the table. One by one, Ge's family members—some are brothers, some are cousins—emerge from down a hallway and greet Henry and Du. They are rich by Guiyu standards, bosses in a business that pollutes their own home, their own mother, and then—and only then—the rest of this small village. Henry introduces me as the son of a U.S. scrap man, and their eyes open wide.

"I want to know the price of some stuff," I tell them, and open my bag. I place each of my five cell phones onto the sofa, along with the rest of the old electronics I've brought along. The young men reach for the devices, turn them in their hands, one by one, and discuss the kinds of chips they contain, the amount of gold that they might hold.

"See, they know the chips in the phones," Henry says with wonder. "Some of the chips maybe they can sell for ten or twelve dollars. But for you, they just pay for the scrap value. Unbelievable how they know all of the chips just by looking at the phone model."

"The phones aren't reusable?"

"Of course not!" He laughs. "They're five years old. Who wants them? Even in Africa, the market isn't so good for these anymore." Even in Africa, where the living standards are often lower than the poor villages that supply labor to Guiyu, they want to upgrade to something better. It's the modern mindset, the source of Guiyu's scrap.

The youngest of the six men tosses my old Samsung phone up and down in his hand and says that the whole lot—all of the phones—will go for around $16 per kilogram.

Next, the group turns to the old PDA. I don't expect much: after all, PDAs haven't had a new market segment in years. Reuse is impossible. They pass it back and forth, turn it over in their hands, pull the plastic cover off and try to pry away the back. There's an argument over the monitor, and then a youngish family member announces the price is around $3,200 per ton. He tosses it onto the sofa, where it makes a leathery thump against the pleather.

The cords are easy: $1,000 per ton, and they'll be shipped off to a wire stripper (not a burner) for separation into copper and plastic. My old Dell's power pack goes for twice that—$2,000 per ton—mostly because it can be reused (possibly marketed to someone in the U.S. via eBay who lost one).

Finally, there's my old Dell laptop.

"The screen works?" the young one asks.

"Yes."

"That can be reused. Maybe worth two hundred yuan [around U.S. $31 on the day of my visit]."

"What'll happen to it?"

"Maybe put it in a new laptop."

It turns out that the PC itself, and its Intel Celeron chip, is too old to be sold for reuse—either in Guiyu, or up in the Shanghai used computer market where this journey started. Still, due to its young age and potential for other reusable parts, the laptop is very much "high grade" by the Guiyu method of evaluating waste. "So probably it goes for around thirty thousand yuan per ton." That's around $4,761 on the day that I visit.

At that, we're done, and there's nothing left to say. They look at me, and then they look at Henry. "Do you want to take them with you?"

I look at all my old phones spread out on the sofa. I remember calling my grandmother on those, telling her what I saw in China's scrapyards. It's just like the old days, I'd tell her. You'd recognize everything. "I don't think so," I answer.

"We'll recycle them for you, then."

I hesitate. Soon, maybe later today, my old phones will be downstairs in that fisherman's cage, awaiting an acid bath that'll turn them into gold and the sweet stench that chokes this dusty town. But that won't be the end. Soon after, that gold and the other raw materials will be sold to a factory that transforms them into new things—smartphones, computers, and the other accessories of daily life.

I reach for a toothpick and stick it into a piece of the watermelon that Ge's mother cut up for me. The fruit is full and juicy, sweet on my tongue, and just maybe filled with the poisons of Guiyu. Who knows? Maybe some of those poisons once carried the voices of friends and relatives. I take a bite, and it's sweet. I look around the room at everyone laughing, catching up as a family, happy. They don't seem to care about the toxic work done downstairs in the courtyard, but maybe they shouldn't. After all, could they live like this, as a family, if they were still farmers? I don't know, and I'm not sure it's polite to ask. Instead, I have another question. "Is Guiyu becoming cleaner? Now that the burning is inside."

Ge shakes his head with a regretful frown. "Not better. Bigger."

"Bigger?"

"There's more scrap coming these days. Most of it's from China. Less and less from the United States. The Chinese scrap isn't as good as the U.S. scrap, so it can't be reused as much."

"How much is from the United States these days?"

"Less than half. Most of it's Chinese."

For years, activists and the media have portrayed Guiyu as the natural outcome of foreign rapaciousness. To shut down the injustice, they implied, scrap recyclers just had to stop shipping computers to China. In retrospect, it was a simplistic message, even before China became the world's largest consumer of computers. Today, though, it's beyond simplistic: it's willfully ignorant. Guiyu won't be shut down because it stops receiving computers from abroad. It'll be shut down if and when China manages to implement an environmentally secure electronics collection and recycling system for all of the electronic scrap being generated within its own borders. As of 2013, the Chinese government is funding pilot programs, research, and—in Guiyu, for example—upgrades. Still, environmentally secure electronics recycling is far from a priority in a place that lacks access to clean air, water, and—in many rural areas— proper childhood nutrition. Right or wrong, for many Chinese— especially in Guiyu—electronics recycling is a route to prosperity that might allow them to afford those bigger problems.

Ge has something that he wants us to see before we leave town. So as the day dims, we drive through Guiyu's villages, past narrow lanes with sign after sign advertising "IC," and three-story buildings where smoke stacks convey vaporized acids up the siding and into the blue skies. Finally, we come to a stop on the outskirts of one of the villages. Outside are a few older buildings from Guiyu's prescrap era, single-story and sturdy, beneath gently sloping tiled roofs.

In the midst of this crumbling history is a single-story temple, hung with red lanterns and decorated with faded ceramic gods and birds. Before the Communists came to power and destroyed the institutions of old China, temples like this one were places where wealthy family clans memorialized their histories, projected their wealth, and above all pre- served continuity with the past. Alas, for the Communists they were a threat, a gathering place for organized powers that might challenge the new powers, and they were destroyed, mostly, during the Cultural Revolution. Those that survived were in far-flung places, as Guiyu once was.

Ge leads us out of the van and through the open door. It's an ornate and impeccable space, hung with intricate plastic lamps, tables covered in red silk, and new brass candlesticks. A couple of old men sit around, drinking tea, watching an old color television. Ge nods at them: they're members of the clan, elders deserving of respect, if not something better to do with their time.

Ge stands proud as he walks us to the front of the hall and an altar of small wooden tablets inscribed with the names of ancestors. "The temple is two hundred years old," he tells us. "But my family has roots dating back to the Song dynasty. The current clan dates back four hundred years."

The beams of this temple also date back two hundred years, to China's last period of widespread prosperity, the early Qing dynasty. But the rest of it—walls to plastic lamps, paintings to stone plaques—dates to this most recent, high-tech-driven period of prosperity. That's how it's always been in China: clan temples reflect family and national fortunes. For now, at least, Ge's clan is enjoying scrap-driven prosperity.

We wander outside and up a narrow dirt path that winds between the abandoned homes that once constituted an old village. The doors are wooden and red, the walls thick cement. Down one lane I see laundry hanging from a line between buildings, and near another lane I smell old urine. There's no scent of burning electronics, here, no stench of wealth.

We ascend a hill that looks down on those old roofs, rising and falling, gently curving, the old China that I never knew, the one so poor that people were willing to trash their lush fields to grow rich recycling someone else's computers. To the right, at the bottom of the hill, five-story buildings rise from old farm fields. They have balconies and big windows and—according to Ge—contain entire family clans that used to live across from each other in these lanes. There's running water in those buildings, and electricity that doesn't cut out constantly. Those clans, Ge tells me, paid for those buildings by recycling imported electronics.

I want to ask Ge what his ancestors might think about what's happened to Guiyu. Would they celebrate the wealth that's bought electricity, plumbing, and a refurbished clan temple? Or would they mourn rivers and canals so polluted that Guiyu's water supply must be trucked into town daily? Was it all worth it? As I try to formulate a polite, and respectful, means of asking these questions, he asks if I'd mind snapping a photo of him and Henry with the two villages—the abandoned old one, and the new high-rise one—in the background. I agree, and as his chest puffs out to pose, I decide that I have my answers.

CHAPTER 12

The Coin Tower

Seven weeks after I moved to China in September 2002, I received a call from James Li, a Chinese-American scrap man who I met when I was working with my family's scrap business. James was different from the other Chinese buyers. He liked talking about things other than scrap, and he took a sincere interest in me, and my family (and especially my late grandmother). When he heard that I was moving to Shanghai for a few months of freelance work at the beginning of my post-scrapyard career (a "premature midlife crisis," is how one friend characterized it), he offered to show me around, including—critically—some Chinese scrapyards.

It was a generous offer. The list of stories available to an American writer with no China experience and no Chinese language is limited. What little journalism had been done on the Chinese scrap industry at that point had focused on the informal, polluting sector dominated by small workshops in places like Guiyu. But here's the thing: even though the small informal sector is an important piece of China's scrap industry, I was fairly certain that it couldn't handle all of the scrap that U.S. trade statistics suggested was moving into China in 2002. And that's what I wanted to see: the large-scale scrapyards hidden behind gates, walls, and other layers of security.

Which is where James comes in.

* * *

The yard James has invited me to see belongs to the Sigma Group, and my first impression as we drive through the gates is of the two ominous shrub-size cast iron busts of Lenin and Stalin perched on both sides of the entrance to the plant. This is what I had in mind!

"You should ask him about those," James says.

"The busts?" I look at James carefully: he's short and a little stocky. In his late forties, he has the squinty smile of a mischievous child. "Who should I ask?"

"The owner. Mr. Huang."

James is an old friend, a warm and friendly guy with the easiest and biggest laugh of any scrap man I've ever met. But he also has a penchant for mischief that sometimes—I think—isn't good for business, though his considerable success in the industry has proved that instinct wrong. "Really," he says, boiling over with laughter. "I'd like to know, too."

We park in front of the short office building and are met in the lobby by a moderately tall, very dapper man in his early fifties. He's dressed in a black blazer, gray tie, and khakis, and he exudes the blunt charisma of someone who takes for granted that he's a born leader of men. "Hey, how are you?" he says in a deep baritone.

James introduces himself—I was under the mistaken impression that he and Tony have met before—and then me.

I'm not really sure what to say, so I do the natural thing and follow up on James's suggestion. "I noticed your Lenin bust at the gate," I tell him.

"Huh? Those things? They came out of the scrap. You can take them if you have a way. They weigh a lot." He smiles a cockeyed smile, and so do we. But we aren't here to collect Communist souvenirs; we're here to see the scrapyard, and Tony is happy to oblige, in large part, I think, because James is keen to do business with him. He gives us hard hats, and we walk outside and around the office.

It's raining lightly, wiping clean two very large piles of shredded scrap perhaps one hundred feet long and five feet high. But this scrap isn't steel. Rather, it's the 2 percent of a shredded car that's copper, zinc, aluminum, magnesium, and other nonferrous (that is, not iron or steel) metals that can't be pulled off by a shredder magnet. In your average one-ton Volkswagen Beetle, that 2 percent adds up to forty pounds of metal—metal valued in the dollars per pound; if you're an automobile shredding company, that's thousands of pounds of metal and hundreds of thousands of dollars—if not more—per month.

Just ask Tony Huang: in 2002 Sigma imported around 100,000 *tons* of shredded mixed scrap metals, 90 percent of which originated in the United States. After sorting them into constituent metals, Huang then turned around and exported the aluminum to Japan, almost exclusively. In fact, he told me, he was the largest supplier of aluminum scrap to the resource-hungry Japanese automobile industry that year. It doesn't stop: ten years later, in 2012, Sigma Group accounted for 40 percent of all Chinese recycled aluminum exports (that is, new aluminum made from scrap).

Scattered around the outdoor aluminum piles are slight figures in teal jumpsuits and surgical caps and masks, shoveling fist-size hunks of crumpled scrap into yellow wheelbarrows—and then wandering off with the loads. Despite the baggy uniforms and masks, it's obvious to me that I'm looking at women: they're too slight, especially their shoulders, to be men. We followed several toward a four-story warehouse capped by a gentle peaked roof and draped in windows. Even fifty feet away, the sound that issues from it is audible and distinct: like a powerful typhoon rain—a rain, in fact, more powerful than the one in which we're walking. It's sharp and metallic, an airier version of the static between analog television stations. We stop at a loading bay door, and what I see shakes me.

Hundreds of slight teal figures crouch on a floor covered in shredded scrap, silently sorting the various metals into plastic bins, each fragment the equivalent of a raindrop in creating the sound of a metallic downpour. As I step tentatively into the space, I'm dumbfounded by the scale of it. The room stretches hundreds of feet, with both sides covered in dark shredded scrap metal that flows from the floor, up onto tables, back onto the floor, over and over. A narrow aisle runs through the middle of the room, dividing one river of metal from the other, and teal-colored workers walk down it with wheelbarrows full of metal that they either dump for sorting or take away, fully sorted.

I step away from Tony and James and wander among the women. From up close, it's even more obvious that they are women: their gloves are too big for their hands, long locks of hair occasionally fall out of surgical caps, and here and there I can see mascara. What I also notice is that they're young: there are no lines around those eyes, no middle-aged fat on those cheeks. These young women are in the throes of first jobs, first adventures from home.

I stop beside a cluster of four teal workers crouched in front of a three-inch-high pile of metal splashed across the floor. Their hands move quickly, without pause: grab a piece of metal, toss it into a bin; grab another piece of metal, toss it into another bin. It's a mechanical movement, perfectly rhythmic, and lacking any doubt or uncertainty. I stand there unable to distinguish between the metals that they're clearly sorting and think: This is what it would look like if I were asked to randomly toss pieces of metal into bins and make it look like I knew what I was doing.

One bin holds grimy red metal that is obviously copper; another, pieces of wire. But the others are less distinguishable. A very full bin, I'm later told, holds aluminum; another, less full, stainless steel. But to my eyes, frankly, they both look gray and metallic, nothing more, totally indistinguishable.

"They have to train for a full month," explains Tony. "They learn to tell the metal by how it feels and looks. They don't make many mistakes."

"Why are they all women?" I ask.

"Women are more precise and patient. Men aren't good at this kind of work."

There are 800 of them working at Sigma, but only—I'd guess—150 in this warehouse. I kneel beside two of them and try to make eye contact as they work. But they won't look at me. I aim my camera at their intensely focused eyes, but that doesn't draw a reaction, either. As my flash bursts, they don't flinch but rather appear to focus more fiercely on the metal, continuing to work through the pile. I stand up and notice more teal-garbed women arriving with more yellow wheelbarrows full of more unsorted metal—beyond sufficient to replace what has just been sorted. It feels endless, the one time in my life that I can apply the adjective Sisyphean and not feel like I'm being melodramatic.

Tony tells me that he pays the sorters around $100 per month, plus room and board. It's a good salary in 2002, and much better than what could be earned on the farms from which most of Sigma's—and China's—factory labor has migrated. There they'd work for subsistence wages without any promise of a future. I'm not sure what kind of future is available at Sigma, but at least the workers earn enough cash to save up for one (China's migrant workers are notorious for saving in excess of half their salaries). Or, rather, their families back home can do the saving: like migrants across China, Sigma's workers send their salaries home to support parents, children, and siblings. In fact, according to my interviews, most workers who live and work at places like Sigma don't keep more than perhaps 30 percent of their salaries for themselves. What they do for that money doesn't look like the sort of thing that I'd want to spend my time doing. Rather, it looks like factory work: tedious, boring, unfulfilling, the sort of work where you spend your time calculating how much extra you'll make if you can manage to get another ten minutes on your punch card.

Tony seems to read my face. "You think they prefer to spend all day planting rice in the fields? At least at my place they only work eight hours and rest on the weekend."

"Really?"

"They work more than eight hours, and they get tired and don't sort so good. If they don't sort so good, my quality drops."

I watch gloves skimming over metal, flipping fragments of aluminum into bins. It's amazing to me that anybody could last more than eight minutes at this kind of work, much less eight hours, *and* do it for $100 per month, plus room and board. But spread out before me are 150 women who seem to think it is worth it. Nobody's forced them to come here; they could've stayed home, wherever that might be.

Tony leads us briefly into another warehouse where towering boxes that look like nothing so much as giant, steel-clad fireplaces take the edge off the autumn chill. As we stand there, the metal door to one slowly rises, lifted by a chain that reveals a hellish orange fire and—I'm told—molten American automobile scrap sorted by the women in the other warehouse. There aren't many workers in the furnace room, and—I note—they are all men, dressed in blue. Perhaps, I think, melting scrap is a less precise business than sorting it. The poor reputation of Chinese metal in 2002 certainly suggests that might be the case.

Tony draws my attention to the right, the far end of the warehouse, where silvery aluminum ingots are stacked atop each other in bundles that rise four feet off the floor. There are dozens of bundles, thousands of ingots. I run my hands over them, and snap a photo of one of the labels:

SIGMA METALS
MADE IN CHINA
CONSIGNEE: TOYOTA

MATERIAL: ADC-12
DESTINATION: TOMAKOMAI
LOT NO: 1021K2-44
NET WEIGHT: 506 KGS

In a week or two that bundle of ingots will likely arrive in Japan, where soon after, it'll be melted and cast into engine blocks and other parts to be bolted into new cars. Some of these cars will remain in Japan, and some will be exported to the United States, where, with any luck, they'll remain on the road for ten to twelve years. Then, like the cars that preceded and literally made them, they too will be sold and eventually shredded, packed into containers, and shipped back to China, where the cycle will repeat, endlessly.

That's the idea; or at least it was, back in 2002.

Nine years later, and I'm watching a dump truck tip eighteen tons of what looks like a reddish pile of dirt, foam, and trash onto the ground at Huron Valley Steel Corporation's complex in Belleville, Michigan. That pile joins some twenty other piles, each perhaps the height of a pickup truck, with circumferences ranging from fifteen to twenty feet. Together, from a distance, they look like a miniaturized, trashed-up version of the Rocky Mountains as viewed from 30,000 feet.

I'm standing next to David Wallace, a senior VP at Huron Valley, and Jack Noe, quality control manager for the plant. The three of us walk up to one of the piles and pause. On first impression, I think: this is precisely the sort of dark, dirty, foamy crap that belongs in a landfill. In scrap industry parlance, it's known as shredded nonferrous, or SNF, and it's everything that comes out of a shredder that isn't steel. Among other items it contains shredded upholstery, dashboards, rubber gaskets, the coins stuck between seats, and all those McDonald's wrappers under the seats. However, it also includes—buried within it—all that metal I saw at Shanghai Sigma.

"So what do you guys think the metal recovery is?" Jack asks, referring to the percentage of the SNF that's recoverable as metal. He's in his mid-forties, but in his gray hoodie and hard hat those round cheeks suggest a much younger and competitive man. He steps forward and pushes a boot into the squishy mass. Depending on the source of the SNF, it can contain between 3 percent and 95 percent metal (Huron Valley doesn't disclose the average), making this a guessing game that favors the experienced scrap man. "I'm gonna say thirty-five percent," Jack guesses.

David purses his lips and stuffs his hands into the pockets of his suede coat. He's a lawyer by training, and his eyes narrow at the material as if he's evaluating it for evidence. "It's hard to say with all of that foam." He sighs. "I'll say twenty-five percent."

"Adam?"

I have no idea. I step forward and press one of my new work boots into the squish. "Twenty percent," I guess.

"Hold on," Jack says. "Let me get the map."

As Jack jogs back to the office, I poke some of the loose fragments of plastic with my boot. It sure doesn't look valuable, and I'm not alone in that opinion. Back in the late 1950s and early 1960s, when the Prolers and Newells first ran their shredders, they landfilled SNF without a second thought. The goal was to extract steel, after all, and that could be done with a magnet. But how was anyone supposed to extract all those nonmagnetic metals from the foam, much less sort them from each other?

In Detroit, Leonard Fritz, Huron Valley's founder, was asking the same question. He'd grown up salvaging dumps, and in the years leading up to World War II he'd made a fortune in the profession. When he returned from overseas service, he rebuilt that business, and in 1963 he and Huron Valley started shredding cars. From the beginning, however, he was troubled by the sight of good but mixed-up metals (the SNF) being carted off to landfills. His treasure-hunting instincts activated, Fritz and the company began looking into methods for extracting all of the SNF that it—and shredders across the United States—were generating.

Today, Huron Valley is the world's largest recycler of SNF. It operates SNF plants in Belleville, Michigan, and Anniston, Alabama, where in 2011 it recycled 775 million *pounds* of shredded nonferrous for automobile shredders all over North America. But that's nothing: in 2007, the most profitable year in the history of the U.S. scrap industry, the year during which the last of the eighty-year backlog of abandoned American automobiles was cleared, they received over *1 billion pounds* for processing.

Why Huron Valley, and not somebody else?

The answer is largely hidden behind the walls of several large but rather nondescript steel-clad buildings that contain some of the most advanced recycling technology ever developed. For precision, at least, the only real competition is what I saw in Shanghai: hundreds of human hands, connected to well-trained eyes.

Jack rushes back with a clipboard, flipping through the sheets of paper. "Thirty-five percent," he says triumphantly, and shows us the figure printed cleanly. The company that sold the SNF to Huron Valley is a regular customer, and—like most shredders—it produces a fairly consistent SNF.

I glance at that printed "35 percent," back at the pile, and do a calculation in my head: that's maybe 12,000 pounds of metal in there (a small percentage of which is worth more than $3 per pound). Or, thinking of it differently: if a shredded car is 20 percent SNF, and each of the piles weighs 18 tons or so, then that pickup-truck-high pile of 35 percent metal is roughly equivalent to seventy Volkswagen Beetles minus the steel. I press my foot against the pile again; it still gives like a sponge. "Doesn't feel like thirty-five percent," I tell Jack and David.

"Don't worry," Jack tells me with a salesman's confident smile. "It is. We check." That's an understatement: they extract it, weigh it, pay for it, and then sell it all over the world.

I follow David and Jack up a set of metal stairs that crisscross beside a conveyor lifting SNF into one of the monumental, nondescript buildings where the material is sorted, and at the top I put the lens cap on my camera. What I'm about to see is highly proprietary, and though some of the techniques and technologies within it are almost a century old, and well known (in principle, at least) in the global scrap processing industry, how they are assembled and made to work together is not. Photos would be tantamount to providing blueprints to competitors unable to equal Huron Valley's precision.

Inside, I need a moment for my eyes to adjust to the dim light, and when they do, my immediate impression is that I've been ushered into a nefarious flume ride. Chutes convey rushing water up, down, and all around a colossal enclosed space, threaded by catwalks and dark, boxy contraptions that shake the scaffolding, and the air.

Yet as I stare across this shadowy space, I'm most struck by how lonely it is. There's nobody in here. In the distance I see a young man in a hard hat moving along a catwalk—but that's it. This isn't Sigma, with its army of women. Rather, this muscular, indecipherable machinery runs on its own, like a movie playing to an empty house. All things considered, it feels kind of timeless, as if those chutes of rushing water were carved by nature, not engineers.

Nonetheless, the point of this building isn't to awe, but rather to separate shredded metal from shredded foam, plastic, and other nonmetallic pieces of a pulverized automobile. The physics of how this happens are deceptively simple.

Consider for a moment, a common egg. If you place it in a bowl of fresh water, it sinks. But if you add enough salt to that water, the water becomes heavier than the egg, and the egg floats. In the mid-1960s Ron Dalton, an engineer who first went to work for Leonard Fritz in 1957, wondered what you'd need to add to water to make various metals float. The idea, he told me in an interview, came from reading a book that described how floatation was used to separate minerals in the mining industry. There, miners used "heavy media"—the industrial equivalent of table salt—to increase the weight of water, and thus float away crushed rock from crushed minerals. It seemed logical that the concept could be extended to scrap metal.

Today, Dalton will be the first to admit that it wasn't nearly as simple as he'd initially thought. To start, it takes something much more expensive than common table salt to float the aluminum in a pile of SNF. Nonetheless, after several years of tinkering, Huron Valley's first "flotation" plant (purchased in part from an old iron mine in northern Minnesota) began operating in 1969. The goal in those days, as now, was to float the trash off the metal, and then to float the light metal (that is, aluminum) off the "the heavies"—copper, brass, and zinc.

On the catwalk, we walk parallel to one of the rushing chutes of water in which media is used to float nonmetallics from metallics. Sure enough, the surface churns, and bits of rubber and dirt float to the top while below, I'm assured, the various metals roll along. Eventually, at a point in the process that's proprietary, the metal is diverted in one direction, and the floating trash, in another. The metal stream is mostly pure by that point, with scattered bits of plastic, rubber, and glass, and it all speeds down conveyors that dead-end into chutes and bins.

At the end of those conveyors something curious happens. The metal fragments don't fall so much as they seem to leap into the air and deposit themselves into chutes and bins they wouldn't reach if they just relied on the momentum of the conveyors. Meanwhile, the remaining trash—the rubber and the plastic—falls away harmlessly.

The device that makes this curious separation possible is called an eddy current. Thomas Edison developed and patented the first one in the 1880s, and it's most assuredly the case that he didn't foresee its application to shredded automobiles. Still, he'd immediately recognize what's going on. The spinning rotor that drives the conveyor is packed with magnets. As the magnets spin, they create a kind of magnetic field around the nonmagnetic metal fragments (such as aluminum and copper, which are not ordinarily magnetic) that approach the rotor. When those fragments reach the rotor, the field repels them just as two magnets repel each other when similar poles are placed end to end. The practical result is this: pieces of metal are ejected from the remaining mix of conveyed trash.

Jack leads David and me down a stairway to the damp plant floor, where Jack points to a hopper filled with the light media's final product: a plastic- and rubber-free mess of wires, crumpled copper, fragments of red, gray, and silver metal, and little bits of gears, brackets, and other parts. The bin is weighed, that weight is reported back to the customer who sent it, and payment is made. Just like that, the bin now belongs to Huron Valley, and it's up to them to figure out how precisely to recycle those mixed metals.

One method would be simple: just ship it to Asia and let somebody—or many somebodies—hand-sort it into pure metals that can be fed into a furnace.

"The Chinese would buy this in a second," Jack tells me. "They love it." They do, indeed: over the last fifteen years Chinese companies like Sigma in Shanghai have imported mixed metals like these by the millions of pounds per year, hand-sorted them using low-cost labor, and sold the cleaned-up metal to raw-material-hungry manufacturers in China and around Asia.

But export isn't always the best payday—especially if you have the technology to compete with hand sorters. After all, in Detroit, and around North America, manufacturers are eager to get their hands on good-quality aluminum scrap to fashion into products ranging from car parts to wire. If an American scrap company has the means to separate aluminum from other metals—and Huron Valley does—then that company has a ready and lucrative customer base in North America. Huron Valley isn't necessarily the future of recycling—they've been using technology to sort metals for decades—but they offer one model (albeit, a proprietary and highly capital-intensive model) for those who want an alternative to more labor-intensive methods of recycling.

Jack, David, and I follow a loader filled with mixed metals next door to the heavy media plant, where aluminum is separated from the other metals. It's a dark, relatively smaller space filled with what feels like an equal number of chutes, ladders, and eddy currents, giving it a cramped, slightly claustrophobic feeling. The crash and clatter of metal on metal fills my ears; the low hum of machinery—eddy currents to conveyors—shake my bones. This too is a flotation plant, but rather than floating trash above metal, this plant floats lightweight aluminum above the heavier metals (including copper, brass, zinc, and stainless steel). As we pause atop a catwalk, I see gray fragments of aluminum literally floating atop frothy water, while down below are what Huron Valley refers to as "the heavies."

But that's just the start of a complex process. I look up, and a conveyor rises rapidly with wet fragments of metal on its surface; I look to my right, and there are more conveyors, more flumelike channels, shunting back and forth like a busy Los Angeles highway interchange. Behind them are car-size boxes that contain eddy currents that not only separate trash from metal but, through careful tweaking, play an important role in separating metal from metal. The principle used is straightforward: different types and sizes of metal are thrown different distances by a magnetic field. If you place bins at the correct distances from the eddy current, you should be able to collect, say, fragments of aluminum less than 20 millimeters in size and pieces of stainless steel greater than 40 millimeters in size, to offer two slightly fanciful examples.

But really, the eddy currents aren't even close to being the most interesting equipment and process here.

David walks me over to a four-foot-tall metal bin positioned just beneath a bathroom-size box he calls the "coin tower." From the coin tower, high-speed round projectiles are shot into the bin at irregular intervals less than a second apart. I step forward to get a better look, and Jack taps my shoulder and suggests I be careful. "You don't want to get hit in the eye," he says.

So I lean over carefully, and there, piled a foot high against the sides of the metal bin, are beat-up U.S. quarters, dimes, nickels, and pennies. Next to the filling bin is another bin, a full bin of coins that fell from the pockets of Americans who had more pressing matters than loose change. According to Jack, an average junked U.S. automobile contains $1.65 in loose change when it's shredded. If that's right—and from what I see, I believe that it must be—then the 14 million cars scrapped in good years (good for automobile recyclers, at least) in the United States contain within them more than $20 million in cash just waiting to be recovered. Understandably, Huron Valley isn't interested in revealing just how much money they recover from U.S. automobiles (they have a deal whereby they return the currency to the U.S. Treasury for a percentage of the original value), but David is willing to note that the coin recovery system has "paid for itself."

It occurs to me that Huron Valley has happened upon the most brilliant of businesses: one whose product is money itself! That is, rather than make something that needs to be marketed for money, Huron Valley just makes money.

I'm taken up to the coin tower to see precisely how they accomplish this lucrative task, and it's awesome—but it's also something that the company prefers that I not describe. Suffice it to say that they have the ability to identify things that look like coins on a conveyor covered in scrap, and then shoot them off the conveyor using precisely aimed bursts of air. In all my years covering scrap, there is nothing I've enjoyed watching more.

Still, for all of its gee-whiz precision, sorting coins from scrap is a side business compared to the primary purpose of the heavy media plant: sorting aluminum from "the heavies." Jack and David walk me out of the damp room and into an airy space between warehouses, where against one wall an imposing dull gray drift of metal fragments climbs maybe ten feet toward the ceiling. Individually, the fragments are the size of padlocks, and as diverse in their battered shapes as snowflakes in a winter drift.

"That's the Twitch," Jack tells me.

Twitch.

By the ISRI specifications, "Twitch" is what you call shredded aluminum that's been separated at a flotation plant of the sort operated by Huron Valley. It's supposed to be 95 percent aluminum (and aluminum alloys), and as I kick through it with my work boot, it sure looks that way. David, however, reminds me that Huron Valley's Twitch actually approaches an incredible 99 percent aluminum. That might not sound like a big difference—95 percent to 99 percent—but in the metal business, it's huge. It allows Huron Valley's aluminum to be used in a much wider range of applications, and—significantly—it allows Huron Valley to charge a premium for it, too.

I drag a few of the lightweight scraps across the floor with my steel-toed boot. Soon, perhaps even today, they'll be loaded into a truck and driven twenty miles north to Huron Valley's furnace in River Rouge, not far from the site of Henry Ford's Reincarnation Department. There it'll be melted into new aluminum that Huron Valley will sell to companies across North America—including, as it happens, automobile manufacturers. In short order, perhaps in less than a week, all of that Twitch—all that scrapped-in-America, recycled-in-America Twitch—will be refashioned into engines, transmissions, wheels, and other essential bits of the mobile American experience.

None of this happens out of good intentions. The American automobile industry buys Huron Valley's recycled aluminum because it's just as good as virgin aluminum, and it's priced competitively. But the green effects of Huron Valley's profitable intentions are undeniable: recycled aluminum is made using roughly 8 percent of the energy needed to make new aluminum, and it doesn't require a bauxite mine (producing one pound of virgin aluminum requires mining roughly five pounds of bauxite ore). That savings goes directly into the pockets of those who demand aluminum for their products, and those who do the recycling.

I glance to my right, toward an open loading door and the waiting piles of yet-to-be-processed SNF outside. Forty years ago they would've been landfilled and forgotten. Today, Huron Valley not only extracts metal from SNF shipped to it from across North America, it's also occasionally engaged in the business of mining landfills where SNF was dumped before flotation and other separation techniques were developed.

For Leonard Fritz, the octogenarian founder of Huron Valley, those mined dumps invoke something deep. As a nine-year-old boy he dug in Detroit's city dumps to earn enough for school clothes. Early one summer afternoon, at the end of a two-hour interview with me in his office, he grows reflective. "See, you're born here," he says behind his desk and the amber lenses that he wears high atop the bridge of his nose. "You're digging and grubbing." He pauses for just a beat to consider his words.

"We started our grubbing in the city dumps at nine years old. And you always look for something somebody threw away."

A few months later, and I'm having breakfast with Frank Coleman, Huron Valley's vice president in charge of nonferrous sales, at a scrap recycling convention in Singapore. Frank is in his late fifties, fit and trim, with the wiry body of a man at least two decades his junior. The son of a millwright, Frank worked himself up from being a hand sorter on one of the earliest separation lines at Huron Valley (in 1971, no less) to an office job where he spends his days talking to scrap buyers around the world interested in Huron Valley's "heavies."

He wears a polo shirt and slacks, and aside from the obvious strength wound tightly into those broad shoulders, he comes off as little different from the hundreds of other scrap traders roaming this hotel. To me, at least, he's just the warm, funny (scathingly so) Frank Coleman who, over the last few years, has become one of my favorite scrap men, anywhere. He's also reflective: in a few weeks he'll have completed his fortieth year in the scrap industry.

As he eats his American-style breakfast, Frank tells me that there are two things he's really enjoyed about the scrap business: the people, and watching the scrap evolve. Back in the early 1970s, he tells me, the metal content of a car was very different. Today it might be 75 percent lightweight aluminum. But back then it was perhaps 65 percent heavies, and 65 percent of that was very heavy zinc. "We were shredding cars that were maybe late fifties, mid-late fifties, to early sixties, you know," he recalls between his eggs. "So if you can picture, these things were just heavy. Big heavy Dodge, lot of chrome, the mirrors were all zinc. Anything that was chrome on a car was made out of zinc. Even the radio knobs."

So long as Americans had big, heavy, shiny cars loaded with zinc, Huron Valley was one of the world's leading recyclers of zinc. The means to recover that zinc was to take the heavies—zinc and all—and heat them until the zinc melted and could be separated from the other, unmolten metals. Among the customers for the recovered zinc (much of it in dust form) were automobile manufacturers who used it as an antirust compound on car bodies.

But that was by no means the most notable—or historical—use of Huron Valley's millions and millions of pounds of recycled zinc over the years. In 1982, when the U.S. government decided to replace the copper penny with a 95-percent-zinc penny, it was Huron Valley that supplied it with 4 million pounds of the metal.

Think about that the next time you come across a 1982 penny in your pocket: you might be holding the recycled remains of your grandfather's chrome-plated hot rod.

In any case, packing so much zinc into a car was all fine and good and shiny in an era of cheap, plentiful fuel. But the oil shocks of the 1970s caused Americans to start thinking about fuel efficiency and lighter—if not smaller—cars. It usually takes about a decade for a car to go from assembly line to scrapyard, and sure enough, in the mid-1980s—a decade after the oil shocks—Huron Valley started noticing a big change in the SNF running through its plants. "We see the zinc being replaced, phasing out, being replaced by aluminum," Frank tells me. "Now all car door cranks, if you even have them anymore, are plastic. You have no more side vents. Side vents are a thing of the past. Getting into the eighties here, you see the change, all of that evolving, the scrap was evolving. The zinc was being replaced by aluminum and/or plastic."

The shift was profound for the U.S. scrap industry, and especially for Huron Valley, which had built a significant part of its business on Detroit's voracious appetite for zinc. But it was by no means the most profound change. That was happening in Asia, where growing economies were beginning to emerge and seek out sources of metal to use in infrastructure and low-cost goods for export back to the United States and Europe. Some, like China, had the means to open mines and smelting operations. But why open a mine, when the United States and Europe have an excess of scrap metal just waiting to be melted down?

"Frank? Frank!"

We both turn around to see Melissa Tsai, founder and general manager of Green Finix, a very large Taiwanese broker of scrap metal into China. She's carrying a leather attaché and a heavy shoulder bag, and she reaches out for his shoulder with affectionate familiarity. Melissa has been doing business with Huron Valley for years; in fact, I met her during a casual dinner with the company in Los Angeles a few years ago.

"Oh, sorry, Adam," she says with a smile. "I just have to ask Frank a question."

Frank has a gently amused smile on his face, as if he knows what's coming. "Aren't we having dinner tomorrow, Melissa?"

"Yes," she answers with a slightly frantic inflection and a glance at me. "But I want to know if I can get some heavies. I need heavies."

Frank chuckles, and I excuse myself. We're having dinner later that night, and I'm not helping anybody if I'm sitting there while he and Melissa try to negotiate a deal to move all those shredded American automobiles to Asia.

In 1986, after an intensive and expensive research program, Huron Valley opened a facility that used a mind-boggling combination of eddy currents and image recognition devices to turn a bin of mixed heavies into individual bins of pure copper, brass, stainless steel, and zinc. The impact on Huron Valley's business was transformative: between 1986 and 1995, the volume of metal that the company recovered from SNF increased almost four times, to well over 100 million pounds per year. All of that was accomplished without a single hand sorter at a sorting table.

The scrap-metal business is a catty, highly competitive industry where even the best of friends will badmouth the quality of what the other produces in his or her scrapyard. But nobody—*nobody*—has ever claimed to surpass Huron Valley's ability to sort the heavies.

Or, more precisely, nobody else has Huron Valley's ability to sort heavies mechanically. Hand sorting, however, is another matter. In the 1970s Taiwanese scrap processors taught their then-low-cost employees how to hand sort, and in the 1980s some of those same Taiwanese exported that know-how to China with the expectation that China would soon have a voracious need for metal to build infrastructure and cheap products to export to the United States.

But there was something else going on, as well.

Chinese and Indian scrap-metal importers are—among other things—highly averse to paying taxes. So consider: would you rather pay a 17 percent tax on something that's worthless, or pay 17 percent on 40,000 pounds of something worth, say, $1 per pound? As it happens, a pile of mixed-up brass, copper, zinc, and stainless steel is worth much less than individual piles of the various component metals (reflecting the cost of sorting it, to begin). And from the perspective of an overworked, undereducated, corrupt Chinese customs officer in 1990, that mixed pile looks, well, almost worthless (Chinese customs officers have become better educated since then, though arguably they are just as corrupt). If you're looking to avoid a hefty assessment on your imported scrap metal, you're better off importing mixed metal than the pure stuff. Then, when it arrives at your Chinese warehouse, you assign a team of $15-per-day laborers to sort it out. That's a cost, for sure, but less than what you'd be paying in taxes if you imported it as separated metal.

In other words, if the Chinese government ever decides that it'd like to put an end to the hand sorting of scrap metal, the quickest route to accomplishing that goal would be to abolish import duties on mixed metals.

According to David Wallace, in the mid-1990s Huron Valley started taking orders from Chinese and Indian importers that went a little something like this: "A container of copper, a container of zinc, a container of brass. And before you send it over to us, would you please mix it back together?" Of course, from the perspective of Huron Valley, this raised a worthy question: why bother separating the heavies at all? So in most cases they didn't bother.

At the same time that Chinese demand for metal was growing in the 1990s, the rising cost of complying with environmental rules was making North American copper refineries less and less competitive with low-cost Chinese competitors. Not only did the North Americans fail to operate as cheaply as Chinese companies, they no longer could afford to pay as much as the Chinese for scrap. As a result, American copper scrap that would have once remained home was heading east, further hobbling the once proud American copper industry. By 2000, in fact, only one copper refinery was left in North America—a situation that persists. Not coincidentally, in 2000 Huron Valley shut down its North American heavies sorting line, and in 2003 it shut down the line it had recently set up in Europe. From now on, heavies would head east, and the aluminum would stay home.

Selling mixed metals isn't as easy as it sounds. It requires somebody who understands, intimately, not only what comes out of the sortation lines but what goes in. Frank Coleman not only had that knowledge, he had knowledge of what it took to build those sortation systems—because he'd been with many of them since the foundations were laid.

When Frank walks out to Huron Valley's scrapyard, he's all business: he wears a white work smock and a hard hat and carries a cigarillo in the side of his mouth. To my eyes he looks like a surgeon in search of a bar fight. He lights that cigarillo, and instead of turning left, to the media plant, he turns to the right and a row of concrete storage bays big enough to hold cars. Instead of cars, though, they hold the shredded remains of cars. Some piles are gray in color; some are reddish in color; and some are in between.

It's a warm August day, and we're joined by Jack Noe, who clearly enjoys Frank's company and knowledge. But again, it's all business: Frank stops in front of a reddish-looking pile filled with wire fragments, crumpled pieces of copper, and lots of silvery and gray bits of other metals. He takes a draw off his cigarillo. "Where's all that manganese shit coming from?"

I double-take. I've spent a lifetime around guys like Frank, but I've never come across somebody who could spot manganese in a pile of metal fragments no bigger than broken pretzels. Truth be told, I wouldn't know a piece of manganese if it was floating in my soup.

Jack doesn't share my surprise at Frank's keen eye—in part because Jack, as the man now in charge of those piles, might be more up-to-date on what's in them than Frank. He kicks at the pile, then turns to me. "Sometimes there might be a problem with the heavy media plant letting too much of something through," he explains.

Frank nods and continues walking down the line. He's in quality control mode now. "What's this?" he asks, pointing his cigar at another bay.

"That's the new product," Jack answers. "The aluminum and magnesium."

I look down the bays of metal, one after the other. Each holds a pile of what were once automobiles. Hard as I try to imagine the cars that this rubble once was, I can't. It's like standing in a supermarket meat section, staring at a package of hamburger and trying to imagine cows.

A few months later, I'm back in Foshan, in search of where all of those American automobiles go after they leave Huron Valley for sorting. As I've learned over the years, there are many reasons that Foshan, China, became the beating heart of China's mixed scrap-metal trade. Its culture is entrepreneurial, its customs agents are notoriously corrupt, and—as important as anything—there once were lots of poor farmers looking for cash wages to improve their lives (the farmers are now rich; the laborers are now migrants hoping to become like the farmers). But perhaps the most important reason—the one that overrides all of the others—is simply that Foshan has nice weather.

Think of it this way: Would you rather try to differentiate between a piece of copper and a piece of brass while wearing gloves, or bare-handed? In Tianjin, a northern Chinese copper scrap hub where temperatures drop well below zero, the workers must do it with gloves, and that means their work is less precise during the city's long, cold winter. But here in Foshan, where the weather is Florida-like (if with a lot more smog), they sort bare-handed, and with precision, all year long. For nearby manufacturers in search of precisely sorted aluminum, copper, and other metals to melt into new products, the choice between all-year precision and seven-month-per-year precision isn't a choice at all. They'll take Foshan's metal, any day.

I'm riding through Foshan's back lanes in the back seat of an SUV owned by my friend Terry Ng (who happens to be a good and steady customer of Huron Valley). He's a soft-spoken scrap man of modest height in his early thirties, nothing like the leathered characters who constitute much of the business around here. In his loose jeans, un-tucked but pressed black button-down shirt, and expensive designer glasses, he looks more like a semi-successful real estate salesman than a guy capable of importing millions of dollars of scrap per month.

He has a bright smile as we drive and talk. And why not? When I met him in late 2008, his family's scrap import company—Ding Fung Limited—was bringing in perhaps 150 shipping containers per month of mixed metals. The math isn't hard: that's around 6 million pounds of metals priced in the dollars per pound. Now, three years later, he tells me that the company has grown to a two-hundred-container-a-month business.

I'd smile, too.

Our first stop is the family's newest yard (it operates under the name Junlong Metal Recycling), located at the end of a dusty road in what feels like the outskirts of Foshan. I step out of the car into a mostly empty space enclosed by a concrete fence. At the far end, beneath a roof, perhaps fifty workers are busy squatting over piles of mixed metals and rubber buckets. Like their counterparts at Sigma in Shanghai, they're primarily looking for aluminum that can be sold to a local melting fa-cility. "And they sell it to Toyota," Terry tells me.

"For use in Chinese cars?"

"I think so." He laughs. "I don't know."

It wouldn't surprise me. In 2002, when I visited Sigma for the first time, the United States was still the world's largest automobile consumer, and Japan was the biggest exporter of cars to that market. So to a certain extent, the American scrap metal that was exported to Shanghai was eventually going to return to the United States. But today China is the world's largest automobile consumer, and its biggest automobile manufacturers have built factories in towns within a short driving distance of here. The American scrap metal that feeds Terry's yard isn't making a round-trip home. More and more, it's staying here, in China, where the country's new middle-class drivers need it for their cars.

As we walk the yard, I notice that the workers aren't the young girls who worked at Sigma in 2002. Instead, I see faces with years and middle-aged paunch on them. Here and there, I also see men. "Young people don't want to do this work anymore." Terry sighs. "They want better kinds of jobs." Ironically, it's already a better kind of job: wages at Junlong, and across Foshan, have crept up to $400 per month. Most of the company's workers don't live in dorms anymore. Rather, they own bicycles and commute to apartments that they rent or own.

Terry walks me into the two-story building that houses, on its first floor, six green plastic picnic-style tables and benches. Spread out over them are thermoses and plastic bags containing lunches. At least from the doorway, it looks perfectly normal, like a small factory lunchroom anywhere.

But there's little time to dally. Terry has an appointment later in the afternoon, and I want to see the heavies plant.

Fifteen minutes later Terry's SUV rolls slowly through the gate of Junlong's heavies-sorting plant. To the right is a one-story office where, the last time I was here, Randy Goodman and I tried to set Terry up with a scrap man's daughter from across town. And to the left is the heavies-sorting operation.

It's the size of a hockey rink—maybe a bit bigger—and it's filled with waist-high hills of mixed metals. In the valleys, perhaps a hundred workers on their haunches sort white bags filled with scrap, filling plastic bins with the various metals—copper, brass, zinc, and stainless—that make up "the heavies."

Here, too, the workers are older, I notice. They're also better paid: around $500 per month. But this makes sense. The pieces of metal are smaller, and the work is more difficult. I lean down beside three women who—I think—must be in their mid-thirties at the earliest. Unlike younger women at other scrap-sorting companies, these ladies giggle at my presence. But when I point my camera at them, they become shy and shield their faces.

Instead, I take pictures of their hands, and the scrap. It's minuscule. Fragments of wire no longer than two inches are intermingled with postage-stamp-size fragments of silver metal. The workers here sort fast— separating silver from silver from red from wire. But they aren't nearly as fast as their sisters in Shanghai who work with bigger pieces of metal.

I'm told that this is the biggest heavies-sorting plant in Foshan— which means that it must be one of the biggest heavies-sorting plants in the world. When I ask Terry if that's true, he just shrugs. The business was built not by him but by his father and other family members, and he's careful to defer to them. For now, he's just doing his best to grow what he's been given. "My father says that the business is now ours," he says. "His working days are done."

That business is going to change, and Terry knows it. Maybe in a few years, once the price of Chinese scrap labor has quadrupled again, Terry will want the kind of automation that his father undercut in the 1990s. That day may be coming sooner than anyone expected, in fact. The rising price of Chinese labor has government regulators concerned that China may not be able to compete effectively for scrap overseas, so they're subsidizing the development of technology to reduce the need for hand sorters. Can China compete against Huron Valley if the competition is machine versus machine? Probably, if only because demand for raw materials in China is so high, and companies are willing to pay whatever is necessary to get them. So long as the demand persists, so will the importers.

CHAPTER 13

Hot Metal Flows

After lunch, Saturday, November 8, 2008, I went for a stroll behind the China World Hotel in Beijing. A brisk wind that day carried dust from the Mongolian hinterlands, and mixed it up with the swirl of dirt that levitates off the city's countless construction sites. My eyes stung, and I decided to finish the walk in the gym.

On the way back I spotted a loading dock where perhaps a half dozen wiry scrap-metal peddlers, with bicycles and knapsacks, were busy haggling with a dock manager over the hotel's empty aluminum beer and soda cans. Both sides had good reason to haggle hard, and I paused to watch.

In the six weeks since the September 2008 collapse of Lehman Brothers, and the onset of the global financial crisis, the price of aluminum had fallen by more than half. This was nothing unusual: as a raw material, scrap is highly sensitive to perturbations in the global economy. If manufacturers aren't manufacturing, the scrap dealers are among the first to feel it. Indeed, scrap prices began to slip as much as six months ahead of the Lehman crash. Likewise, if an economy is starting to grow, scrap prices tend to be among the first rising indicators (when Alan Greenspan was the chairman of the Federal Reserve Bank, scrap prices were among his favorite economic indicators). At my family's scrapyard, we always felt we knew the direction the economic breeze was blowing months ahead of time.

What was different about 2008 was how fast it happened, and how far prices fell—in the case of some grades of steel, they dropped 80 percent in a matter of weeks. Nobody in the industry had experienced anything like it—just as nobody in the industry had experienced anything like the decade-long boom in scrap prices that preceded it. Both phenomena were driven by globalization—by the ability of scrap exporters to meet the insatiable demands of scrap importers. When the demand evaporated in the face of fear, the bottom fell out of the markets.

The scruffy scrap peddlers behind the China World probably had never heard of Lehman, but they knew damn well that the street price of Coke cans in Beijing was half what it had been a few weeks before. As a result, either the China World's dock manager would need to drop his price, or he and the hotel would be stuck with smelly trash bags filled with cans in need of an unlikely buyer.

To my surprise, the dock manager held firm against the frustrated attempts of the peddlers. Perhaps, in his stubbornness, he believed that the price would eventually rebound, that all of the recently abandoned construction sites in Beijing—the ones kicking up all of the dust—would resume, need window frames, and drive up the price of cans again. In November 2008, though, it didn't look like a very good bet.

Aluminum cans weren't the only recyclable items trending into worthlessness on that brisk November day. On a nearby corner, piles of cardboard—once a highly sought commodity among Beijing's cagey scrap peddlers—sat in the early winter sun, awaiting a taker. But nobody was taking: the economic crisis had sapped the will of *Americans* to buy anything, and since most anything that Americans purchase is packaged in cardboard, the price of used cardboard, too, had plummeted. By day's end, the China World's cardboard, once worth a day's wage to a scrap peddler, would be on its way to a landfill or an incinerator, and nobody—not a scrap peddler, an environmentalist, or a dock manager hoping for some cash on the side—could do a thing about it.

I was at the China World with an invitation to observe an unusual summit: at 2:00 P.M., some of the world's biggest and wealthiest scrap-metal buyers and sellers were to gather at a long table in Function Room 12, a stuffy narrow space on the hotel's lower level, in order to settle their differences over broken contracts that were threatening the globalized trade in metal recyclables. There were at least two billionaires among them, and several who—if they offered their companies publicly—could attain that status. The others, ranging in age from their late thirties to their seventies, and from countries around the globe, China to Haiti, the United States to Nigeria, were wealthy beyond the imaginations of the ragged, wiry scrap peddlers haggling at the loading dock—despite the fact that both parties were in the business of buying other people's waste for cheap, and finding a way to sell it for more. The differences are profound, but in many ways they come down to one simple factor: the peddlers sell locally, and the titans sell globally.

Not that any of the mostly paunchy men in Function Room 12 were thinking in such esoteric terms in November 2008. Rather, most of them were either angry or frightened, having watched as global prices in some scrap metal, paper, and plastic declined by as much as 80 percent in the space of the previous few weeks. The decline devastated Chinese recyclers, in particular, many of whom had had built up large inventories of American and European scrap metal in anticipation of a perpetual, Chinese-driven run-up in the price of raw materials. China's economic development had been viewed as inevitable, and as more and more of the country took part in that development, the price of metals could only go up.

So they bought more and more scrap metal, often on the basis of a 20 percent down payment on 40,000-pound shipping containers that might contain scrap worth $100,000 or more. But when in the fall of 2008 the price of scrap metals went into a free fall, many of these traders made a cold and calculating decision: it's better to hold losses at 20 percent than to pay for an even bigger loss—perhaps as much as 50 percent. And so, starting in September, hundreds of Chinese scrap buyers refused to pay the outstanding balance for the containers of scrap metal that had already sailed from ports in Europe and the United States. They broke their contracts. By early November Chinese ports were jammed up with *thousands* of devalued containers of American scrap metal that nobody wanted to claim as their own, much less pay for. Only a week before, I'd been to the port of Ningbo in southern China, where containers were piled like children's blocks while outraged port officials ranted about the problem to any scrap man—foreign or Chinese—who dared to visit.

The wealthy foreign exporters gathered in Function Room 12 were there to make money, and the wealthy Chinese scrap traders, many of whom had plenty of money in reserve, wanted to renegotiate the terms of their contracts to save money. Friendships and partnerships built over a decade were frayed, if not permanently damaged, and in this absence of commerce and comity—in the absence of a market for developed world recyclables like telephone cables, washing machine motors, and shredded American automobiles—scrap metal and cardboard began to pile up in municipal recycling warehouses across America and Europe.

By early December the recyclables that Americans and Europeans had diligently sorted into blue and green bins were now worth no more than the abandoned containers of scrap metal and paper at Chinese ports. Desperate recycling managers in the United States and Europe looked at warehouses overflowing with cardboard, and they did the only thing they could: they sent it to landfills and incinerators.

The meeting in Function Room 12 was convened by the China Nonferrous Metals Industry Association's Recycling Metal Branch (CMRA), a mouthful of an organization that serves as a trade association, government agency, and scrap-trading business—all at once. Not many people in China, much less outside it, had ever heard of the CMRA. But if you participated in the multibillion-dollar transpacific scrap trade over the last decade, you couldn't risk ignorance about who wields power in China's scrap industry. The CMRA's power extended over the entire industry, from policy making to enforcement; if something went wrong at a port, the CMRA heard about it and resolved the issue—discreetly; when the highest levels of the Chinese government decided that scrap should be a "strategic industry" on the level of oil, they asked the CMRA to draft a plan to make it happen. The CMRA is far from all-powerful, but in China, where true power is elusive, the CMRA was and is an undeniable, tactile nexus, the tongue-twister agency with a hand on one-third of the metal going into China's world-beating manufacturing sector. That's power.

The timing for the meeting was fortuitous and last-minute: most of the world's major foreign scrap traders were in Beijing for the CMRA's Secondary Metals International Forum, an annual conference then in its seventh edition. Most years, it was a lavish affair where dry, poorly attended speeches on policy are afterthoughts to banquets, ballrooms, and, above all, business. At past events the lobby, and the lobby bar, were jammed with buyers and sellers trading business cards and prices. Away from the hotel and the conference hall, Chinese scrap traders splurged on outrageously expensive and elaborate meals (I once attended a banquet that featured a large cooked alligator in the middle of the table) and fine-boned prostitutes, all in hope of convincing a handful of foreign scrap suppliers that they—and nobody else—should be the ones to buy their scrap.

But the financial crisis put an abrupt end to the fun. In 2008, unlike, say, 2007, there were no Taiwanese scrap brokers in the lobby whispering "Do you want a girl?" to any male with a white face, a delegate badge, and a shipping container of old washing machine motors to sell; there were no invitations to lobster sashimi banquets with free-flowing everything; there were no expense-paid weekend trips to Hainan, "China's Hawaii," with yet more "girls." Instead, in the midst of the market free fall and recent confirmed reports that a British scrap trader had been kidnapped by a government-owned scrap company in southern China over a commercial dispute, all of that was replaced by a tense, paranoid atmosphere. I saw one Chinese buyer turn and jog away from a U.S. supplier he'd recently stiffed; I saw another Chinese buyer with whom I had an acquaintance show up with a new company name—and a new personal name on his delegate badge. Several participants arrived with personal bodyguards, just in case. By the time the "special meeting" convened in Function Room 12, foreign suppliers were speaking only with foreign suppliers, and Chinese buyers were speaking only with Chinese buyers.

I was in Function Room 12 by invitation, in my role as a journalist. But in the eyes of the CMRA, I was no ordinary reporter, no mere member of the Dow Jones or Bloomberg financial staff. Rather, for the last six years I'd been the only journalist in the world who covered the Asian scrap trade for two of the most important periodicals in the global scrap business, *Scrap* and *Recycling International*. That had provided me unmatched access to the men in Function Room 12. I'd visited their factories, met their children, enjoyed their alligator banquets, sang karaoke with them, and when appropriate and useful, exchanged information. In part, that's why I had a seat at the table, and not in one of the several uncomfortable chairs pressed against the wall, facing the table. Scrap men, and Chinese officials generally, don't like foreign reporters; but when they do, they treat them very well. I know for a fact that the CMRA was not always thrilled with my reports on the Chinese scrap industry, but they'd come to trust that I—as the only reporter around devoted to their industry—was at least accurate. From that basic understanding a relationship, and access, developed.

The mood in Function Room 12 was formal and controlled—a rude surprise to the Americans, in particular, many of whom naively (despite their millions of dollars of trade with China) expected a freewheeling, open-ended discussion that led to a resolution (in their favor, of course). But the long table, with Chinese mostly on one side and foreign exporters mostly on the other, was devised in part to create an illusion that comity had been restored. In other words, it attempted to suggest that— despite breaking thousands of contracts with their American suppliers— the Chinese importers could still sit down, chat, and even trade with their cheated suppliers. And you know what? Two years later, the volume of scrap being exported to China from the United States and European Union was almost back to the levels of 2008.

But I'm not sure the exporters saw rapprochement as inevitable in November 2008.

Rather than open the Function Room 12 meeting to discussion, the CMRA had organized a set of carefully worded addresses by officials and economists that amounted to little more than a declaration that the Chinese side, too, was facing problems. There were no apologies, or even acknowledgments, of what so many Chinese importers had done to their trading partners, and this did nothing to calm the nerves of exporters facing millions of dollars in losses.

The public statements and behavior of certain Chinese business owners didn't help, either. Luo Guo Jun, the general manager of Ningbo Jintian Copper, a massive privately held enterprise that specializes in recycling imported copper, rose from his seat and asked for a microphone. Over the last several weeks, rumors had been rife that Jintian had reneged on millions of dollars' worth of contracts with people in Function Room 12, causing significant financial losses. To stem the losses on its own side, Jintian was reportedly shutting down factories and seeking help from the government. Luo, a proud, cigarette-thin and cigarette-stained middle-aged man whose hair was just starting to disappear, looked nervous as he stood with a sickly smile and spent several long minutes recounting his company's proud twenty-year history and good reputation. "I am familiar with rumors that we failed to honor our contracts," he began softly, speaking through a translator. "Since we were established, we've never failed to pay our bills in an up or down market." As he waited for the translator, he forced a smile. But that smile faded quickly as his words were greeted with groans, sighs, and angry chatter.

Bob Garino, then the director of commodities for the Institute of Scrap Recycling Industries, reached for the microphone and, turning directly to Luo, snapped, "What you described as a rumor is fact."

Two seats down Salam Sharif, the handsome, bearded, and always well-dressed scion and president of Sharif Metals, a Middle Eastern scrap-metal empire that stretches from Jordan through the Gulf States, the Saudi peninsula, Yemen, and Somalia, was rocking back and forth in his chair, seething. Sharif, whose business thrives on recycling what's wasted by the Gulf's wealthy consumers, its ever-booming oil industry, and the bullet casings and other remains of the wars that occur within and around it (so-called conflict scrap, and surely somebody must do it, so why not Sharif?), grabbed the microphone and reminded the Chinese that the Middle East imports large volumes of Chinese finished products—like iPods. "From the belief that what comes around, goes around," he warned, via a flustered translator, "this must be resolved." In the months following the meeting Sharif would form a Middle East Recycler's Association that, rumor has it, blacklists not only Chinese importers who break contracts but also Middle Eastern scrap traders who do business with them.

Next, Robert Stein, a burly former philosophy graduate student turned VP at Alter Trading Company, one of North America's largest scrap recycling firms (and by extension one of North America's largest exporters by volume), took the microphone and turned his eyes to "Lester" Huang, a major northern Chinese scrap importer who had proudly showed me Stein's photo on his wall when I visited his offices a year earlier. "It's very unfortunate that the trust that has developed between buyers in this country and suppliers abroad has evaporated in a month," he said grimly. "You can't make metal products out of air."

The threat contained an undeniable truth: China's manufacturers had become addicted to American and European scrap metal over the last two decades, and if—somehow—it were cut off, they'd be looking for a new source of raw materials. The Chinese government, if not its scrap-metal importers, surely could see the wisdom in not further angering its exporters.

It was nearly 4:30 when the Function Room 12 meeting broke up. Nothing had been resolved—at least, not on the surface—despite hours of interminable explanations interrupted by angry rants that often outpaced the abilities of the beleaguered translators. It got so bad that by the end, Chinese and foreigners alike were speaking so fast and with such volume that they could only be understood by their respective language cohorts. The translators were left behind, helpless. But the message was clear on both sides: the foreigners wanted to be paid, and the Chinese wanted to renegotiate. Nobody was happy. It was a wasted afternoon. The only solution was time.

A year earlier, the global recycling trade had been the ultimate seller's market, and anybody who could bid higher than the last guy was assumed to be an honest business partner. Demand was so high, and competition so fierce, that American and European scrap suppliers chose their customers from hundreds of suitors. It had even come to pass, unsurprisingly, that some of the more pragmatic Chinese replaced their young male sales reps with attractive young females, some of whom—it was said—obligingly slept with the customers in hope of obtaining a few extra shipping containers of aluminum.

I walked into an empty corridor and took a seat on a plush velvet bench. Nearby, a Chinese with a Hong Kong accent was speaking softly with two European scrap traders. From what I could overhear, the Hong Kong trader wanted to buy the abandoned containers that the Europeans' original customers had left at port. The Europeans didn't look happy, but they didn't have much of a choice, either. After all, it was better to take a big loss than a total one. The Hong Kong buyers weren't being charitable. Sooner or later, they knew, American consumers would start buying—and recycling!—again, and those Chinese who bought scrap low, now, would sell high later for the purpose of supplying Americans with more things to buy and throw away. And if the American economy didn't recover, well, the Chinese economy was growing, with plenty of new middle-class consumers interested in buying cars, personal computers, and iPods made from American scrap.

The Chinese, more than the Americans and Europeans, seemed to recognize that the stalled recycling trade was temporary. They knew that the Americans, for all of their anger at broken contracts, would trade with China again, whether they liked it or not. "There's no manufacturing in America now," scoffed James Li, my longtime friend, the man who introduced me to Sigma, and a businessman who now owned yards in China and the United States. "So Americans will have to sell their scrap to somebody. And China will be the biggest one for buying again when the economic crisis ends."

In 2013 the United States exported more than 42.8 million metric tons of scrap metal, paper, rubber, and plastic to 90 countries around the world. The value of that export trade was $23 billion, according to U.S. government statistics. But that's almost certainly an understatement that doesn't account for all of the scrap that's smuggled to avoid taxes and regulators. Whatever the figure, the volume of scrap exported from the United States and other developed, wealthy nations has grown exponentially since China's economy began to rev in the late 1990s, and more than 900 percent—in the case of aluminum scrap—between 2000 and 2011.

I felt that demand, personally, when in the early 2000s I began attending Chinese scrap-metal conferences like those run by the CMRA. These events typically draw several hundred Chinese scrap traders and processors, hardly any of whom are interested in presentations regarding government scrap import policies. Rather, they're interested in meeting—and soliciting—the relatively small handful of foreign scrap-metal exporters who would attend these events for networking purposes. Generally speaking, the exporters are not difficult to spot: among hundreds of Chinese faces, they are usually the paunchy, middle-aged male Caucasian faces. My face was neither paunchy nor middle-aged, but it is Caucasian, and thus it was all but impossible for me to walk through a scrap convention without at least a couple of hungry importers rushing over, business cards and brochures in hand, with a simple pitch: "Can I have your card, please? I'm interested in Birch/Cliff [bits and pieces of copper]."

For a journalist, this sort of interaction had its advantages. Many of these scrap-hungry traders were eager to be featured in the magazines for which I work, and I was eager for sources. But after a few years of source gathering, I'd had enough of their aggressive ways. And so at a 2006 convention I asked a Chinese friend to write, "I have no scrap," at the bottom of my name tag in hope that it might save everybody—me and the traders—time and aggravation.

It didn't work.

CHAPTER 14

Canton

It's fall 2011, and Johnson Zeng is driving precisely five miles per hour above the speed limit on Interstate 77, south through West Virginia. Trucks are passing us, cars are passing us, but Johnson maintains that steady pace. He drove that speed this morning, from Cincinnati to Canton, Ohio, and he's driving that way now to the Carolinas. It's how he always drives. Suddenly he has a question: "Why is Guangdong called Canton?"

I wasn't expecting that.

Historically, "Canton" is the anglicization of Guangzhou, the capitol of Guangdong, Johnson's home province. Likewise, Cantonese is the anglicization of the language spoken in that region. Rather than answer him, though, I wonder aloud about something else: "I wonder if people in Canton call themselves Cantonese."

"No," he says with an earnest shake of the head and a bit of home province pride. "Belongs to Guangdong people."

I turn to him; he's not smiling. I wasn't trying to be funny, and apparently I wasn't. Despite his Canadian citizenship and wandering ways, Johnson knows precisely what he is—Cantonese—and no Ohioan is going to move in on that.

In fact, neither of us has particularly warm feelings for Canton, the Ohio version. Earlier in the day, Johnson had an appointment with a

yard there, and we drove four hours out of our way—a way that was supposed to lead to North Carolina—to make it, only to find that the yard
manager had sold off any and all export scrap the day before. "Yard owners and managers should call and say there's no material available," Johnson tells me in a rare burst of frustration. Then he pauses to reconsider his
words. "But they're busy. That's what I'd do, too. Seller's market."

It's midafternoon, and we have at least six hours of driving ahead of
us before we reach Statesville, North Carolina, home to tomorrow's first
appointment. "Normal day for a Chinese trader," sighs Johnson. "Long
drive, no payoff." That's not exactly true. A few hours earlier, in the
parking lot of the Canton scrapyard, the Cincinnati scrapyard that we'd
visited earlier in the day called to agree on a deal for two containers
worth roughly $70,000—bringing him closer to his goal of spending $1
million this week. "Still, I planned on four [containers]," he admits.

Spend even a few days on the road with a Chinese scrap trader, and
you can't help but start to think that the odds—and the American scrap
dealers—are arrayed against him. Broken appointments, sold-off stock,
and employees who don't bother concealing their racism. Those are just
the affronts. The misdeeds run deeper.

According to Johnson, and other traveling traders, scrap exporters
don't always ship scrap orders as quickly as they promise—especially if
the markets are moving up and they can sell it again for more. "But
when it is falling, then they cannot wait to send the metal to China!"
Johnson laughs ruefully. "Also, when the market is up they like to load
less weight than forty thousand pounds [allowed in a shipping con-
tainer]. When I call to complain, they always say that they cannot risk
being overweight because of highway regulations on how much scrap
can be transported to the rail yard or port. But when the market is
down, there are no highway regulations, and the container always comes
overweight!"

Those aren't new complaints. Joe Chen, president of Tung Tai, repeated them to me as he recalled his days as a traveling scrap buyer in
the 1970s and '80s. "They load the container with not what we bought,"
he told me softly in the back of his Mercedes as it sped through Foshan
in November 2009.

"If we saw this pile—and even if we are there—they push the other
pile into the container."

When I asked Joe why this happened in his traveling salesman days—

whether it was racism, or just the very human assumption that all foreigners are stupid—he answered, "Yes. You are correct."

I feel bad for Joe, but I also know that these tricks aren't the sole provenance of American scrap dealers. Europeans use them; Japanese use
them; Chinese use them. "Games," as most scrap dealers call them, are
part of the business no matter where it's done. Johnson and the other
traveling scrap dealers just have the misfortune of experiencing them in
a foreign country. But trust me, if Johnson traveled the highways of
China, he'd experience much the same.

As Johnson drives, he takes a call from a two-man team of Chinese
buyers in Arizona. They've just closed a deal, and they'd like for him to
arrange the logistics for the containers. It's a new business for Johnson
and his wife, who recently quit her job in Vancouver to work with him
full-time. As that call ends, he starts another to ask that she arrange shipping on the Arizona containers, as well as the ones he closed in Cincinnati. Tonight, at the hotel, he'll forward the information.

Arranging logistics isn't nearly as lucrative a business as trading
scrap, but it involves much less risk. The scrap that Johnson bought in
Cincinnati this morning must be loaded into a container, driven to a
rail yard, and then loaded onto a container vessel chartered to cross the
Pacific. When finally it arrives in Hong Kong, forty-five days will have
passed. The global markets will not have remained static during that
transit: they may have gone up, or they may have crashed. After all, some
markets lost more than half their value in the fall of 2008.

We continue south, dusk falls, and we pass a police car parked behind
a car with a flat tire on the shoulder. "Police are so nice in America."
Johnson sighs. "They stop to help you change a tire." Both of us know
that in China, they might stop and fine you for having a flat.

As the sky darkens, we return to talking about scrap. Higher taxes on
imported scrap are a nuisance, he tells me, but by no means are they an
importer's biggest or most recent problem. That honor belongs to whoever is stealing scrap out of his containers as they move from Hong Kong
to Homer. According to Johnson, and two other exporters to whom I
spoke to subsequently, somebody is skimming scrap off the tops of loads
while containers are under the control of Chinese customs.

The losses can total hundreds or even thousands of dollars, and needless to say, that adds up. Nobody knows precisely who is doing it, or even
where. But theories on how they do it are rife, with the leading candidate

being this one: at customs, mixed loads of scrap are unloaded, assessed, and weighed. "They used to just look and inspect," Johnson tells me. "Now they weigh and steal."

A few weeks later I talk to a very large, well-respected U.S. exporter who tells me that things are actually much worse than Johnson indicated. As an example, he tells me that his company recently "lost" a bale of copper—retail value around $20,000—somewhere between Hong Kong and the mainland, despite the fact that he'd taken precautions, and even paid to have the container weighed in Hong Kong before it was sent to the other side. But that didn't dissuade the thieves.

Later, though, I can't help but wonder: Why would Johnson or this exporter be surprised? Guangdong's scrap trade has always been quasi-legal, at best. If it's tipping into criminality, that shouldn't surprise anyone.

Tonight Johnson drives through the darkness, and we talk about being foreigners in each other's respective countries. Along the way we spot a bright full moon clearing the stars out of the sky. "The full moon is always bigger and better in foreign countries," he tells me. "In China the moon is a positive sign. But in your country it's a bad sign, the full moon is."

"That's true."

"Why?"

I think of werewolves. "I don't know. It looks good tonight, though."

A few weeks later I learn that Johnson was paraphrasing a Chinese aphorism that, roughly translated, reads: "The moon seems fuller in foreign lands than in China." In the traditional meaning, at least, it's a warning against wanderlust and straying from the motherland.

In the morning we check out of our Super 8 and drive less than a mile to L. Gordon Iron & Metal, a venerable scrap processor located on a narrow but long strip of land surrounded by thick green foliage and tall, stately trees. Here Johnson is greeted like an old friend. The ladies who work at the front desk greet him by name—"Hi, Johnson"—in their honeyed accents, having no need to look at the card he places on the counter.

One of them takes us into the cramped, windowless office of Richard Gordon, a fiftyish, obviously type-A fourth-generation member of the family that founded and still owns this century-old scrap business. He's on the phone, and he lifts a finger to indicate he'll be just a minute. As we stand there, Scott Bergmeyer, the tall, thin, and lightly bearded warehouse manager, arrives with orders to show Johnson around the warehouse. Gordon looks busy, so we head out.

"How are your prices today, Johnson?" Scott asks as we cross a driveway into the warehouse.

"Good!" he answers. "Any Chinese buyers stop by this week?"

"You're the second."

I look away and smile. That brief conversation pretty much summarizes three decades of interactions between Chinese and U.S. scrap men: a lot of Chinese chasing a limited supply of expensive metal.

The warehouse is full, and Johnson walks through it snapping photos of giant bales of Christmas tree wire, insulated copper wire, thick black communication cables, metal signboards, and a box of brass machine-shop grindings. At the last item, Johnson pulls out a magnet and runs it over the material, checking whether it's been contaminated with iron. Scott points to a box full of scrapped power drills, and Johnson shakes his head. "When Homer comes." Apparently it takes an expert—somebody who's disassembled a power drill—to know the value of the metal inside it.

"Taxes are going up, prices are going down," Scott drawls.

I look over at him, assuming that he's talking about U.S. taxes. But he's not. He's talking about, of all things, rising Chinese import taxes. Ten years ago nobody in the U.S. scrap industry would've needled a Chinese buyer about customs procedures in Guangdong Province; now even the warehouse managers do. "You keep doing that," he opines, "you won't be able to buy anything."

When we make it back to the office, Richard Gordon is still on the phone, but he gestures toward some chairs. It's a small, cluttered space where two items stick out: the blue-and-white University of North Carolina clock on the wall, and the Carolina Panthers "club seat" sticker on the window. "How much it weigh?" he bellows toward someone in another room (and, presumably, the person on the other line).

"Sixteen thousand pounds," someone calls back.

"Sixteen thousand pounds?" Richard pounds something into the calculator on his desk, pauses at the result, and then looks up. "Twenty-three cents!" Then he returns to his phone conversation.

Meanwhile, Johnson opens a portfolio and starts working on a purchase order. As he writes, he double-checks his BlackBerry for prices on the London Metal Exchange.

"Excuse me?"

We all look up to see one of the ladies from the front desk in the doorway.

"Johnny's on the line for you."

Richard places his current call on hold by punching a button on the phone, then punches Johnny's line and announces, without missing a breath: "Johnny, I'm gonna quote you a big price. Sixty-eight cents." There's a long pause. "He's paying what? Okay, I'll pay you seventy."

It goes like this until Johnson hands Richard his completed purchase order. Richard looks at it briefly, mentions that one of Johnson's competitors has been doing better "on some items," and the prices start moving up. The price of red brass, in particular, is "waaaaaay" too low, but from the way that Johnson starts to slump, I sense that it's not going up. Then something funny happens: Richard tells him he has enough for a "straight load"—that is, a load with nothing but brass.

"Really?" Johnson asks, perking up. "Let me call Homer."

The call lasts twenty seconds, and when it's over Johnson raises his price. To my surprise, he explains why: "It's an easier [customs] declaration," he says. "Lot of trouble with mixed loads lately." In other words, there's much less risk that a container holding a single kind of metal will become lighter en route to China from Hong Kong, presumably because it doesn't need to be unloaded for inspection. From Johnson's point of view, it's worth the extra few cents per pound. The container, I note via a mental calculation, will cost Johnson and Homer around $100,000.

As I listen to Johnson and Richard negotiate, Louis Gordon, a fifth-generation member of the family, stops by to introduce himself, and we start to chat. We're roughly the same age, I think, and definitely of the same background. But unlike me, he actually does business instead of writing about business. Sitting here, amid family that walks up and down a short hallway lined with magazine covers detailing Gordon family commercial triumphs, I'm envious. I would have liked to spend a life in such a business, surrounded by family and "junk."

Then he reminds me of something. "Margins are so tight these days. It's tough."

"So I've heard." I tell him that I've been traveling with Johnson, seeing just how tough it is. And then it occurs to me to ask him a question that wasn't yet a big question back when I worked in the business. "What would happen if you didn't have a China to which you could send your metal?"

He shrugs. "A lot of it would go into the landfill."

I look over at Johnson, negotiating hard for a load of insulated wire. There are two containers at stake, and from what I can tell, he's likely to get them. But if he doesn't, there's no question that another Chinese buyer will.

CHAPTER 15

Ashes to Ashes, Junk to Junk

As I write this last chapter, there's an iPhone 4S sitting on the desk beside me. It's a fine phone, capable of far more than I need, but like high-end consumers everywhere, I'm aware that there's a newer, better phone out there: the iPhone 5. Whether I need the upgraded iPhone or not (and I really don't), I *want* the upgraded iPhone. However, I'm restrained by two considerations. First, it's an expensive device; and second, I'm intimately aware of the environmental costs associated with manufacturing new electronics and disposing of old ones.

Apple doesn't appear to spend much time worrying about complaints regarding its pricing policies, but complaints about its environmental footprint have been a corporate priority for years. Indeed, to its credit, Apple has been at the forefront of technology industry efforts to use fewer, and greener, materials in the construction of electronic devices (all the while making its devices more difficult for individuals to repair and refurbish—a problem I'll address a bit later in this chapter). Even better, Apple is apparently making an effort to refurbish on its own and, when necessary, recycle the products that it manufactures.

Of course, Apple doesn't engage in green practices only for the good of the earth; it also engages in them knowing that an environmentally minded consumer—like me—is much more likely to buy a phone from a company that promises to take back and recycle an old one.

At the moment, the screen on my Sony laptop shows the website for the Apple Recycling Program for the United States. It includes this text: "By participating in the Apple Recycling Program you are helping the environment by extending the useful life of products that have value in the secondary electronics market. You are also ensuring that products that have reached the end of their useful life are recycled in an environmentally responsible manner in North America."

The phrasing is interesting: note how Apple claims to recycle in North America, but gives no details on where it refurbishes and sells refurbished equipment. More likely than not, the refurbishment takes place outside North America (Apple responded to my inquiries, but would not reveal the location of the refurbishment facilities), where labor costs make the work affordable—and customers for lower-cost Apple products are plentiful. I have no problem with that: offering technically oriented employment to people in poor countries is a good thing. Offering refurbished goods to people who can't afford new is an even better thing.

I input the details of my current phone into the Apple website, and it informs me that I'll receive a $215 gift card in exchange for it—$215 that I'm free to apply to an iPhone 5. It's a great deal for me: I'll save money on a new phone with the knowledge that I've behaved in a green, sustainable manner. But is it really such a hot deal for the planet?

The January 2013 issue of the *Journal of Consumer Psychology* contained the results of two experiments that should concern anyone who embraces recycling as a means to preserve natural resources and promote a sustainable lifestyle.

In the first, researchers asked study participants to evaluate a new product—in this case, scissors—by cutting up paper in various, preordained configurations. Half of the study participants did the evaluation in the presence of a trash bin, only, and half did it in the presence of a trash bin *and* a recycling bin. The results were troubling: those who performed the task in the presence of a recycling bin used twice as much paper as those who could only throw their excess paper in a trash bin. "This suggests that the addition of a recycling option can lead to increased resource usage," wrote the authors, Jesse Catlin and Yitong Wang.

The second experiment took place in a more natural setting: a university men's room. For fifteen days, the researchers measured the daily number of paper hand towels tossed into the trash bins positioned next to the sinks. Then they repeated the experiment by adding a recycling bin and "signs indicating that certain campus restrooms were participating in a paper hand towel recycling program and that any used hand towels placed in the bin would be recycled." After 15 days, the researchers ran the data and found that restroom visitors used approximately half a hand towel more when a recycling bin was present than when there was only a trash bin. That may not seem like much, but consider: on an average day, 100 people visited the restroom, meaning that—on average—the recycling bin (and associated signage) likely contributed to the use of an additional 50 paper hand towels per day. Extend that usage out to the 250 business days per year that the restroom is used, and in that one university restroom an additional 12,500 towels would, theoretically, be tossed into the recycling bin, annually!

Isn't recycling supposed to promote conservation and preserve the environment? Why are people using more hand towels if a recycling bin is present? And does this have anything to do with my newfound willingness to buy an iPhone that I don't need to replace my current one? The authors of the study offer a hypothesis: "The increase of consumption found in our study may be partially due to the fact that consumers are well informed that recycling is beneficial to the environment; however, the environmental costs of recycling (e.g., water, energy, etc. used in recycling facilities) are less salient. As such, consumers may focus only on the positive aspects of recycling and see it as a means to assuage negative emotions such as guilt that may be associated with wasting resources and/or as a way to justify increased consumption." Elsewhere in the paper, the authors add: "We believe that the recycling option is more likely to function as a 'get out of jail free card,' which may instead signal to consumers that it is acceptable to consume as long as they recycle the used product."

It's important to note that the authors aren't opposed to recycling. They readily acknowledge the environmental benefits of recycling versus digging up or drilling for new resources. But neither do they believe "simply making recycling options as widespread as possible is the best course of action" for the environment. Rather, it's the third best course of action, after reducing consumption and reusing what's been bought already.

Reduce. Reuse. Recycle.

No doubt the environment would be better off if everyone stopped consuming so much (though people in developing countries without much to consume may take issue with the idea that "preserving the environment" is more important than improving their living conditions via increased consumption). But the likelihood of that is essentially nil, maybe less. Convenience trumps sustainability for most everyone I've met in my travels, American or not. Consider, for example, the cardboard sleeves that we wrap around our Starbucks cups. The one wrapped around the cup on the table next to my computer tells me that it was made from 85 percent postconsumer fiber (that is, recycled paper and cardboard). Should I—a recycling journalist—feel good about that fact? Or should I—and Starbucks—endeavor to stop using paper cups and sleeves altogether, even if it's a little less convenient and comfortable? Couldn't that recyclable cardboard be put to better use somewhere else?

Even those of us who should know better about consuming in excess simply can't help ourselves. In November 2010 I was outside a Hong Kong recycling conference when I ran into Jim Puckett, the environmental activist whose landmark 2002 documentary, *Exporting Harm*, is singly responsible for drawing global attention to Guiyu's e-waste industry. Jim was holding a plastic bag filled with new electronic gadgets, and as my eyes fell upon it, he joked that he'd just "bought some future e-waste."

One day Jim's future e-waste will likely begin its afterlife in an American recycling facility where it'll be shredded. Some of the fragments will remain in the United States, but much of it will end up in Asia, where it'll almost certainly be hand-sorted in a facility that doesn't meet U.S. safety and environment standards. Wherever it goes, and however it's recycled, it'll incur costs.

I leave it to others to write about how and why convenience—and not sustainability—motivates consumers around the world (one theory: consuming is more fun than conserving). Rather, I'll just repeat what I noted at the beginning of this book: between 1960 and 2010, the volume of recyclables that Americans harvested from their homes rose from 5.6 million to 65 million tons. That sounds pretty good until you realize that during the same period the amount of trash generated by those same Americans rose from 81.1 million to 249.9 million tons. Americans—and the rest of the world—are consuming, and recycling, more than ever, *in parallel*. Clearly, increasing the recycling rate isn't going to impact the volume of waste being generated unless we all develop a more nuanced understanding of what recycling can—and can't—do.

When I speak to groups about recycling, the first question is usually: What can we do to improve our recycling rates? I have two answers, the first of which is this: If the goal is conservation, then boosting recycling rates is far less important than reducing the overall volume of waste generated—recyclable or otherwise. Recycling, as I noted earlier in this book, is often just a means of fighting off the garbage man for a little while longer.

Cardboard and paper cannot be recycled infinitely. Depending on the type of paper, the individual fibers can only survive intact for perhaps six or seven trips through the energy-intensive process required to turn them into new boxes and new sheets of paper. Likewise, many plastics can only survive one turn through the recycling process before having to be "down-cycled" into unrecyclable products like plastic lumber for backyard decks.

Metals are a different story. Theoretically, a copper wire can be recycled indefinitely, but that assumes the wire itself is easy to harvest. Extracting copper from a power cable is a relatively easy process; extracting copper from an iPod is quite difficult and usually results in some loss—*especially* when done by recyclers in the developed world who depend on shredders and high-tech to sort the wires from the rest of the materials. However, even relatively simple, well-trod recycling processes, like the one used to recycle an old beer can into a new one, result in some loss of metal along the way, ranging from cans that fall off trucks to metal vaporized in furnaces.

Then there are the things that we'd like to think are recyclable, but simply aren't. Take, for example, an iPhone screen. Glass, in general, is an easily recycled substance that often isn't for one very simple reason: the primary ingredient—sand—is cheap, and thus there's little commercial incentive for a business to seek out and remelt used glass. Of course, an iPhone touch screen isn't made from the same glass as a beer bottle. Rather, it contains a range of so-called rare elements, including indium, a valuable mineral that—at the time I write this chapter—costs more than $200 per pound. Alas, there is no commercially viable means of extracting indium from touch-screen glass, and there is unlikely to be one (the amount of indium in a touch screen doesn't amount to more than a pinch, rendering its extraction a very dubious business indeed). For the foreseeable future, indium—one of the rarest of elements—will likely be mined, used in a single iPhone, and then lost for good.

Nothing—*nothing*—is 100 percent recyclable, and many things, including things we think are recyclable, like iPhone touch screens, are *un*recyclable. Everyone from the local junkyard to Apple to the U.S. government would be doing the planet a very big favor if they stopped implying otherwise, and instead conveyed a more realistic picture of what recycling can and can't do.

Of course, if Apple included that kind of information on the Web page where it explains its recycling program, it might not receive so many old iPhones for recycling, or sell so many new iPhones to sustainably minded consumers like me. It's a point that Jesse Catlin and Yitong Wang, authors of the two recycling experiments in the *Journal of Consumer Psychology*, make in the very last sentence of their paper: "Therefore, an important issue would be to identify ways to nudge consumers toward recycling while also making them aware that recycling is not a perfect solution and that reducing overall consumption is desirable as well."

The purpose of this book is not to provide policy guidance to government officials, or recommendations for those worried about where to drop off their twenty-five-year-old VCRs. However, for those who come to these pages with green intentions, hoping to boost the recycling rates in their communities, I can think of no better advice than the above sentence. As a proud junkyard kid, there's nothing I like seeing more than lots and lots of scrap metal moving in and out of scrapyards. As a self-identified environmentalist, I'd like to see less, especially in rich countries, which generate the most. The best way to achieve that latter goal is to educate consumers that recycling isn't a get-out-of-jail-free card for their consumption.

Nonetheless, if the goal is a *realistic* sustainable future, then it's necessary to take a look at what we can do to lengthen the lives of the products we're going to buy anyway. So my second answer to the question of how we can boost recycling rates is this: Demand that companies start designing products for repair, reuse, and recycling.

Take, for example, the super-thin MacBook Air, a wonder of modern design packed into an aluminum case that's barely bigger than a handful of documents in a manila envelope. At first glance, it would seem to be a sustainable wonder that uses fewer raw materials to do more. But that's just the gloss; the reality is that the MacBook Air's thin profile means that its components—memory chips, solid state drive, and processor—are packed so tightly in the case that there's no room for upgrades (a point driven home by the unusual screws used to hold the case together, thus making home repair even more difficult). Even worse, from the perspective of recycling, the thin profile (and the tightly packed innards) means that the computer is exceptionally difficult to break down into individual components when it comes time to recycle it. In effect, the MacBook Air is a machine built to be shredded, not repaired, upgraded, and reused.

Theoretically, it should be possible to make desirable, thin electronics that are easily repairable, and easier to disassemble and recycle. Traditional PC desktops, for example, allow for the easy switching out of old components for new ones as technology evolves. New memory, a new hard drive, and a new video card: these are installations that anybody with a screwdriver can perform. They save money, and they reduce—not eliminate—the demand for raw materials necessary to make wholly new machines. Finally, when it comes time to recycle, the old modular desktops are easy to break down into their component parts.

For consumers hoping to do something about the growing volume of e-waste piling up, worldwide, a campaign to demand that manufacturers introduce design for recycling principles into new products would go a long way to keeping electronics out of landfills for the long term. At the same time, consumers can encourage the development of reuse by buying refurbished machines themselves. Dell, Apple, and other leading electronics manufacturers market refurbished products with full warranties; the next time you're in the market for a new device, why not consider one of those?

Of course, designing for recycling goes well beyond laptops and mobile phones. Consider the expensive and complicated sensors that a company like Waste Management uses to detect plastics in an automated municipal recycling factory. If a Tide detergent bottle enters one of those systems, and the plastic bottle is covered in a large label made of a substance different from the bottle (paper, or a different kind of plastic), the sensors run the risk of misidentifying the paper or plastic label as the actual packaging material. When that happens, the plastic bottle might end up in a load of paper, or even in the refuse pile. Perhaps it'll be fished out, eventually, and placed into the correct bin for recycling. But even if it ends up in the plastic bin, it's still the case that the paper label will not be recycled; rather, it'll be dissolved in the plastics recycling process.

That's a small loss, but it gets at a bigger problem that's come up in the preceding pages: the more materials that go into a product, the more difficult that product is to recycle.

Consider the Mead Five Star notebook from which I'm typing notes at the moment. It's a simple product: two hundred pages of white paper wedged between a cardboard back cover and a plastic front cover, and bound together by a steel coil. Every one of those items—steel, plastic, cardboard, paper—is recyclable. But who or what will tear the plastic and cardboard covers from the steel coil and paper? If I toss that notebook into an American recycling bin, it'll likely be sensed by a machine at a recycling plant, and diverted to a paper mill (assuming the sensors identified it as paper), where it'll be shredded into its parts. The plastic, most likely, will be mixed with other plastics and landfilled (mixed plastics are not made into new in developed countries), the steel will be sent to a scrapyard, and the paper and cardboard—which, if separated, could be made into new cardboard and new paper—will be turned into a lower-quality recycled paper. That's recycling, for sure, but it's poor recycling: nothing (except perhaps the steel) is recycled as well as it would be if it were separated by hand before the paper mill.

But what if Mead just turned around and designed its Five Star notebooks to be made of paper, cardboard, and—unrealistically—yarn (which can be recycled with paper) to bind it? The cardboard and paper would likely still be mixed together, but there would be no wasted plastic, and the energy wasted on recycling steel could be devoted to something else. It's a simple example, but one that could be extended across a range of products and industries.

The second question that I typically receive when speaking to groups about recycling usually goes something like this: How do we prevent our recycling from going to China and other developing countries, where it's recycled in polluting, unsafe conditions? If the preceding chapters have succeeded at all, it should start to be clear that this question is closely related to the first one ("How do we raise our recycling rates?"). After all, increasing the rates of recycling is a laudable goal, but it won't solve a thing if there aren't people who want to recycle all of the stuff that we harvest in our recycling bins.

For now, and the foreseeable future, the United States will be able to recycle most (perhaps two-thirds) of that harvested material. But there still remain the many millions of tons that American recycling companies can't recycle because American manufacturers (and by extension, American consumers) simply don't have use for additional raw materials at this time. Perhaps at some future date American manufacturing will expand so that it can use all of the Christmas tree lights and other low-grade scraps that flowed into its landfills before China and other developing countries started buying them. But that seems roughly as likely as Americans embracing austerity and rejecting iPhone upgrades.

If the goal is to promote the maximum recycling of what Americans throw away, then American recycling should be allowed to flow to the places where it is most needed. In other words: if Americans want Christmas tree lights, and Chinese companies need copper to supply the wires to make them, then those Chinese companies should be allowed and even encouraged to import used American Christmas tree lights that would otherwise end up in an American landfill. Likewise, if Chinese want to harvest reusable memory chips from broken American computers, then they should be able to import those computers. No doubt, those Christmas tree lights and memory chips won't be recycled in facilities that meet every last health, safety, and environmental standard demanded by the U.S. government. But prohibiting the exports of recycling to developing nations won't do anything to improve those standards, either. Rather, those conditions are only going to change as living standards rise, and the far more pressing issues that face every

developing country—food safety, proper nutrition, and clean water—
are solved first.

Globalization of waste is now a permanent feature of the world economy, no different—and no less permanent—than the globalization of
smartphone manufacture. So long as goods are made in one place, and
consumed and thrown away in another, there will be companies that
specialize in moving that waste to where it's most valued as a raw material. More often than not, those companies belong to what my grandmother called the junk business.

The third and final question I'm asked about recycling usually comes
from friends. It goes like this: "Where should I recycle my *X*?" That *X*
could be an old computer, a bag of newspapers, a box of old wine bottles, a pile of tires in the garage, or an old metal screen door on the back
porch. Sometimes, but not always, there's a follow-up comment: "I really want to do the right thing."

If the items in question are simple—cans, cardboard, newspapers,
plastic bottles, and other home recyclables—I usually suggest finding
the nearest scrapyard. The markets in these kinds of recyclable are so
efficient, and the demand is usually so high, that whatever the entry
point into the global recycling system—a small "We Buy Junk" truck or
a major scrapyard—it's all but guaranteed that the items will end up
recycled by the person or entity that can extract the most value from
them. Better yet, you'll be paid for your recyclables (something that
won't happen if you just leave them for your local government recycling
pickup). The same goes for your car: whether you sell it to a junkyard or
a tow-truck driver, parts of it will eventually end up being shipped to
China, where every last recyclable piece will eventually be recycled.

Electronics and other complex devices offer a different kind of challenge. Functioning but obsolete electronics are high-demand items in
many developing countries, and there are plenty of organizations and
recycling businesses engaged in exporting them. Just poke around the
yellow pages or online to find one. Broken electronics are another matter.
Generally, electronics recyclers in the United States and Europe will test
old equipment, and if it's found to be nonfunctioning, the device will be
sent to a shredder (the economics of repair are difficult). In my opinion,
this is misguided: broken electronics typically can be fixed, but if they

can't, they certainly contain reusable components. But even when they don't, the simple fact is that developing-world recyclers, using hand labor, are able to harvest more recyclable material from a device than a shredder using magnets and eddy currents. For the home recycler, this requires making a choice—developed-world recycling or developing-world recycling, and all that both entail. Hopefully, this book gives you the information necessary to make the right choice for you, the environment, and the hundreds of millions of people who work in the global recycling industry.

Above all, though, I encourage people to think about what it means to recycle, and make smart choices as a consumer before you buy that thing you'll eventually toss out. Recycling is a morally complicated act. For those seeking black-and-white certainty, the local junkyard is a frustrating wash of grays where "thinking green" usually means the dollar bills in the petty cash drawer. But as I noted in the preface, the world is a better, cleaner, and more interesting place for its junkyards. I wouldn't want to live on a planet without them.

On a hot August day, First America Metal in Joliet, Illinois, looks like any American machine shop. It's located at the end of a cul-de-sac, surrounded by a green lawn, and accented by a tall flagpole that flies the American Stars and Stripes. Nothing about the place suggests it's one of the most successful Chinese American–owned and –run scrapyards in the United States. In fact, most people don't realize that scrap men with deep personal connection to China—and they're almost all men—are quietly buying up and running scrapyards across the United States. The motivation isn't hard to discern: they want to cut out the middleman—in this case, the American scrapyard standing between China's raw materials importers and the Americans who toss all that metal and paper into their recycling bins. It's harder than it sounds, as many Chinese scrap companies have already found out, but if it can be done well, the rewards are significant.

This Joliet scrapyard is owned by my old friend James Li, a naturalized America from Hangzhou, China and the man who took me to Shanghai Sigma back in 2002, and he's giving me a tour of his warehouse. Briefly, we pause beside a box of defective home food processors. James picks up one that's been broken apart and shows me the baseball-size motor inside it. The best way to get at the copper is by hand, using a pincer and perhaps a hammer and screwdriver. That's not going to happen in the United States, so—predictably—that box is bound for one of James's scrapyards in China, where it can be done cheaply.

James stops beside another box. "You know what this is?" he asks with a big smile.

I peer into it: there's an oily mix of gray metal shavings of the kind left over when a factory grinds a block of metal into a rounded shape. "No. What are they?"

"Titanium."

Titanium is an expensive, extremely strong, lightweight metal commonly used in aerospace—and golf clubs. A few years ago, while traveling in Taiwan, I visited the island's biggest titanium recycler. It was a memorable visit: I was shown sheets of titanium from which putter heads had been punched out like cookies from dough. The scrap at First America, however, looks more like oily confetti, and I suspect it's not easy to find somebody who wants to buy it. James, however, has the potential to surprise. "Fireworks," he tells me. "Titanium burns white. So you sell them to the fireworks makers. They make white fireworks with them."

I glance in the box again. "Fireworks?" Maybe those shavings were ground off an airplane jet engine part destined for Boeing. Whatever the source, they're now bound for a central China fireworks factory, where they'll be packed into shells destined to turn the sky bright white. "How do you find those fireworks buyers?"

"I know where to look," he answers. "The American scrap guys don't know where to look."

That's true—American scrap guys don't have connections with backcountry Chinese fireworks manufacturers—and it's one reason why James, and not an American junkyard owner, is the natural destination for titanium shavings generated in the United States. After all, successful recycling, or in this case reuse, isn't just a matter of price; it's also a matter of knowing who wants the scrap.

James leads me through the warehouse doorway and into muted offices that could belong to a small real estate company. "The face of the company is American," James explains as we walk the single hallway. "But the office is Chinese."

It is, indeed. The receptionist is a white American; but the small handful of offices behind the receptionist are occupied by Chinese, speaking Chinese. We slip into a conference room where James gives me a seat at the head of a long conference table. He occupies one a few seats down from mine, leans back, one knee pressed against the table, and asks me about my family. I'll be married soon, and he'll be invited, for sure. "What about business?" I ask.

"Not bad," he says. "Always problems, but not bad."

According to James, First America Metal ships 3.2 million pounds of metal per year, making it one of the top five nonagricultural exporters of any product, by volume, from the very agricultural American Midwest. Only 3 to 5 percent of James's metal remains in the United States, where it's mostly sold to auto parts manufacturers. It's a business he believes is endangered. "The U.S. auto companies import a lot of stuff, the parts. So that kills a lot of manufacturing here." In other words: the rise of low-cost Chinese manufacturing in auto parts could push the last of James's scrap to Asia, where demand—and prices—are better.

We talk for more than an hour, and as the conversation tails off, I ask him if he ever feels resentment from his native-born U.S. employees. He's making all of this money, after all, off what nobody in America wants at all. He smiles. "I don't know. I don't know what's in their hearts."

So I ask James if I can interview one.

To my surprise, he agrees. "I'm really curious about this, too. Hold on." He leaves the office and returns a few minutes later with Shane Gilbertson, his sandy-blond thirtyish yard manager. Shane is muscle-bound and handsome; he wears a greasy green T-shirt and a turned-around baseball cap.

"How'd you end up in the scrap business?"

He tells me that he's a college graduate and a former Barnes & Noble clerk. During the 2000s he ran his own financial services business that specialized in mortgages. Then the housing bubble popped. "And you know how that went." He sighs. First America, he tells me, has a lot of guys working for it who, like him, are "happy to have the job," especially in a down economy.

I ask him what he knew about the scrap business before he entered it. Shane smiles and tells me that he's a "former farm boy," a fact that informs everything he is and does. "Here's an example," he says. One afternoon, when he was still a teenager, he and his grandfather took an old hayrack down to "John the Junkman." John the Junkman was not, as Shane describes him, the sort of man who leaves a wealthy, or even a positive, impression. Young Shane certainly wasn't impressed, and he tells me he "made a comment" to his grandfather about John. His grandfather, a man who clearly had some experience in the world, corrected his grandson. "He told me, 'This guy has more money than anybody you know.'"

Times change, and these days I'm pretty sure that James Li is now the guy who has more money than anybody Shane Gilbertson knows. Shane, however, doesn't appear to have any problem with that. When I ask him how he feels about watching all of that good American scrap metal ship out for China, he crosses his arms. "I see it as a flaw in the U.S. economy. I see a lot of the stuff we recycle that could be used here."

"Why isn't it?"

"We have this mentality that we're too good for certain industries. At the same time, we're keeping stuff out of the landfill, and people see us as bad somehow."

Unfortunately, Shane doesn't have much time to talk. It's getting to the end of the day, and he's needed in the warehouse. "We have a lot of organizing to do. I'm still pretty new around here."

Something Shane says sticks with me after he leaves: "I see a lot of the stuff we recycle that could be used here." At first, I think he means that the scrap could be recycled in the United States, rather than exported. But later I realize that Shane was actually getting at something deeper, that he's saying something about what Johnson calls the "big waste country" he crosses in rental cars. There's so much waste, so much promise, and it's all going overseas.

Late July 2012, and I'm in a van with Belly Feng, an overseas sales manager for Hunan Vary, a scrap-metal equipment manufacturer and processor in central China. Belly is an earnest man in his late thirties, perhaps, and he looks precisely as his name suggests: short, round, happy. He is also confident: the very highest levels of China's central government support Hunan Vary's clean recycling program, both politically and financially. Among other notable achievements and projects, Hunan Vary is currently responsible for the design and manufacture of some of the upgraded technologies being installed in the upgraded Guiyu.

Out the window I see the slow-moving Miluo River winding through the gentle vistas of northwest Hunan Province, roughly a thousand kilometers inland from Shanghai. It's a gentle landscape, terraced with fields, lined by small villages being built out into bigger ones as they ride the crest of recent prosperity.

Eventually we turn left to drive down a still-under-construction road, its blacktop stained with the red mud of Hunan Province, until we reach a set of foothills and—on our right—Hunan Vary's ten-acre plot of land. It's dominated by two shopping mall–size warehouses, one of which is abutted by a four-story office building. It's raining, and the mud runs reddish brown down tire tracks and gullies.

As we step out of the van, my eyes are immediately drawn to bluffs built from thousands and thousands of old televisions stacked five and six high, viewed through the open loading door of a warehouse that must sprawl two acres, at least.

Together, we walk through the loading door and wander between the television canyons. Some of the sets are as small as lunchboxes, others are as big as suitcases; the cases are red, white, and mostly brown and black; the screens are unbroken, the dials still turn, and the power cords are still attached. But above all they go on and on, television after television, ashes to ashes, junk to junk.

Belly tells me that many of the sets still work, but nobody in China is interested in owning a twenty-year-old black-and-white television anymore. Even in China's used electronics markets, big color sets are common and cheap.

I watch as two workers in blue uniforms push a squeaky dolly piled with six newly arrived sets down one of the rows. They stop near the end, and together they lift the sets onto the piles.

I feel as if I've just found out where all of China's old televisions go to die. But that's not quite right. According to Belly, this is where *some* of the televisions generated in Changsha and Yueyang, nearby cities of 7 and 5 million people, respectively, come to be recycled. These are the old televisions collected from the 0.92 percent of China's population spread over just those two cities. Elsewhere in China are the old televisions that once belonged to the other 99.08 percent.

Some might call that statistic an environmental problem; others might call it a business opportunity. Here at Hunan Vary, a company supported at the highest levels of the Chinese government, it's both.

Arranged throughout this newly built complex are disassembly lines devoted to the recycling of televisions and their associated wastes. It's a complicated array, but the thing I notice first is that much of the space is devoted to stations where workers disassemble the televisions into their respective components—glass, copper, plastics. Afterward, some of the material is shipped directly to recyclers, and some of it is run through shredders. In simple terms that means Hunan Vary is able to extract and sort much of the nonferrous metal from a computer monitor before it's reduced to fragments, minimizing the need for further sorting after the shredder. Hunan Vary isn't alone in taking this approach: I've seen it in India and other developing nations with low-cost labor that can be thrown at what Americans and other wealthy people shred.

In other words, if—like me—you have a television that you'd like to see recycled in the most environmentally sound manner possible, with the most material harvested from its guts, Hunan Province might very well be the place for it to go.

Is this the future? Will American and European televisions eventually flow to Miluo for green recycling? According to several high-level regulators in Beijing, that's a question that China is close to answering in the affirmative. And why not? If—as seems inevitable—China becomes the world's biggest generator of waste, why shouldn't it then be the biggest recycler, too? If China remains the world's biggest manufacturer, why shouldn't it be the biggest harvester of raw materials from the castoffs of other countries? Why shouldn't it be the capital of Junkyard Planet?

Afterword

In April 2012 I traveled from Shanghai to Las Vegas to attend the annual convention of the Institute of Scrap Recycling Industries. Accompanying me was my fiancée, Christine. She'd never attended a scrap convention before, and I can't say that she was excited. But that wasn't the only reason for her attendance.

A few months earlier Christine's mother informed us that April 18, the Wednesday that fell in the midst of the convention, would be an especially auspicious day in the Chinese calendar. As it happens, it would also be an especially auspicious date in the Jewish calendar, too. So, being ethnically Jewish and ethnically Chinese, we decided that April 18 was the perfect day to be married. Our guest list was small, international, and scrappy: it included a Dutch couple, a Brazilian, two Chinese Americans, and two native-born Americans (all of whom were registered for the convention, of course). The location was a limousine that rolled down Las Vegas Boulevard. Our witness was Kent Kiser, the publisher of *Scrap* magazine.

We loved the event and our guests for many reasons, but among the most important was that they reflected the international diversity of the industry in which I grew up, and now cover. I don't trade scrap, but many of my friends do, and these relationships—these international relationships—are the absolute key to why scrap now flows so smoothly to markets around the world.

But here's the thing: that wasn't always the case.

In the summer of 2011 I spent several days rummaging through the ISRI archives, looking for historical information that would add context to this book. Among the documents that I found most interesting were photos and accounts of the annual dinners of the trade associations that preceded ISRI. The National Association of Waste Material Dealers (NAWMD), for example, typically held its annual banquet at Hotel Astor in New York. A photo I found taken at the 1924 banquet shows a lavish affair that required an entire ballroom, several dozen tables, and several hundred feet of patriotic bunting. But what's most striking, from the perspective of ninety years, is the lack of diversity in the attendees: it was a dinner for white (mostly Jewish) men in tuxedos.

That wasn't just a matter of demographics (though it is certainly the case that the American trade was dominated by male Eastern European immigrants in those days). The annual NAWMD gatherings were officially "stag"—a tradition that persisted into the mid-1980s.

Times have certainly changed.

These days ISRI is an international organization, with international members, and the ISRI convention—all 5,000-plus delegates to it—is an international event. Chinese traders mingle with Indian traders who mingle with African traders, and everybody chases American scrap suppliers. Women are still a minority at the show, just as they are in the industry as a whole, but their numbers are growing, and their influence is spreading (especially in Hong Kong and Los Angeles). No doubt the scrap industry—at the management level, at least—remains a male-dominated industry. But I imagine that in twenty years, that too will have changed.

Nonetheless, I'm aware that people outside the industry—especially in the environmental community—don't view the globalization of the waste and recycling trade with such warm feelings. They view it as outsourcing, dumping, an encouragement to pollute. I understand their concerns: recyclers in developing countries don't generally meet the standards that rich countries impose upon themselves. In some cases they can't afford to improve; in the case of China, and Guiyu, they have the money to improve, but politics and the scale of the problem prevents it. The question is, though: Are they able to conserve more, even if they operate dirtier? Are copper and gold mines better or worse than extracting gold in Guiyu? Is it better to reuse a computer chip in China, or shred it in a warehouse in North America?

In the end, those questions aren't going to be answered by wealthy recyclers in the developed world. Rather, they'll be answered by people in developing countries who need raw materials.

Recycling is better—I won't write "good"—for the environment. But without economics—without supply and demand of raw materials—recycling is nothing more than a meaningless exercise in glorifying garbage. No doubt it's better than throwing something into an incinerator, and worse than fixing something that can be refurbished. It's what you do if you can't bear to see something landfilled. Placing a box or a can or a bottle in a recycling bin doesn't mean you've recycled anything, and it doesn't make you a better, greener person: it just means you've outsourced your problem. Sometimes that outsourcing is near home; and sometimes it's overseas. But wherever it goes, the global market and demand for raw materials is the ultimate arbiter.

Fortunately, if that realization leaves you feeling bad, there's always the alternative: stop buying so much crap in the first place.

As for me and Christine, we don't recycle at home. Rather, we keep all of our plastic bottles, cans, and cardboard in a bin that we give to Wang Qun Ying, our beloved housekeeper (or, in Chinese, *ayi*) of many years. She's an interesting woman, a survivor of China's midcentury revolutionary upheavals who now makes her semiretired living by cleaning up after foreigners like us. The recycling is her twice-per-week bonus, and it's incumbent upon her to get as much as she can for it.

So, twice per week, she leaves our home with a bag or two of what my friends back in the United States would call "recyclables." They go downstairs, seven floors, to the street and the small-time scrap trader who pays for the stuff. We never ask for the money, though once in a while I do ask to know the current price of cans or plastic bottles.

What makes it interesting, from a scrap man's perspective, is when we have something unusual for her to sell. For example, a few weeks after Christine and I returned from Las Vegas, we had a problem with our kitchen sink. The solution involved replacing a cast iron drainage pipe that weighed approximately eight pounds. When I offered it to Wang Ayi, her eyes widened: that's worth some real money. I told her that she could have it, but on the single condition that I accompany her to the scrap trader downstairs and watch the transaction.

At that, Wang Ayi turned to my wife with a concerned look on her face. "If scrap buyers see a foreigner, they won't pay as much money."

"Why not?"

"They don't think you really know the true value."

I laughed. "Then you go without me. Get the best price."

Ten minutes later, she returned with a few yuan in her hand. She shared the price that she was paid, and I promptly looked up the price of cast iron scrap on one of the handful of websites that tracks China's scrap-metal prices. It turned out that she received roughly 30 percent of the market price, and that's not bad. Getting the market price means owning a steel mill and all the volume and problems that such an enterprise entails.

Later that night, I received a phone call from an old friend in a U.S. scrap company. He was calling me with some information for my book, but before we got to it, he had a question: "What do you hear about steel scrap prices these days? We're hearing they're soft."

I glanced at the notebook in which I'd written Wang Ayi's price. "Actually, I was just talking to a Shanghai steel scrap trader today," I told him.

Acknowledgments

In 2002 Kent Kiser, then the editor and now the publisher of *Scrap*, commissioned my first piece of scrap journalism. It worked out, and soon Kent was regularly commissioning China-based stories from me. A few years later Manfred Beck, the publisher of *Recycling International*, also began commissioning stories from me, often in combination with Kent. From the beginning, both publishers gave me license to travel to some of the most godforsaken scrapyards on earth, never questioning the expense, or whether the story was a good idea. It's the sort of support that most journalists dream about but never have. Thanks to Kent and Manfred, I had it, and thus this book was possible.

Sincere thanks to my agent Wendy Sherman, for believing in me, in my writing, and in this project; to Peter Ginna at Bloomsbury Press for embracing this book, and his guidance throughout the process of bringing it to life; and to Pete Beatty of Bloomsbury for knowing precisely what this book needed, and how to get it out of me.

I owe so much to the Institute of Scrap Recycling Industries (ISRI) in Washington, DC. Specific thanks to Robin Wiener and Scott Horne for a wealth of insights and access over the last decade; Bob Garino and Joe Pickard for their knowledge of markets and invaluable help with statistics; and Tom Crane for his work in organizing and sharing the extraordinary trove that is the ISRI archives.

Deepest appreciation to my friends at the China Nonferrous Metals Industry Association's Recycling Metal Branch (CMRA), and very special thanks to the brilliant Ma Hongchang for sharing his deep knowledge of China's scrap industry with me over the years.

Over the last decade I've visited well over a hundred scrapyards around the world, and I'd like to offer thanks to those who gave access, insights, and time in exchange for nothing more than my promise to be fair. I'd like to offer specific thanks to those companies and organizations that provided visits and interviews specifically for this book, as well as companies and people whom I visited over the years, and who are mentioned in its pages. In alphabetical order, they are:

Allied Services Corporation, Alpert & Alpert, Armco Renewable Metals, Cash's Scrap Metal & Iron, Cozzi Partners, DingFung Limited, First America Metal, Freedom Metals, Friends of the Boundary Waters, Fritz Enterprises, General Motors, GJ Steel, Good Point Recycling, Green Finix, Hennepin County, Hunan Vary, Huron Valley Steel Company, Jinsheng Copper, J. Solotken & Company, Junlong Metal Recycling, Jai Varudi Enterprises, Jayesh Impex Pvt. Ltd., L. Gordon Iron & Metal, Leder Bros., Metso Lindeman, Mid-Carolina Recycling, Net Peripheral, Newell Enterprises, OmniSource, Prime Impex, Pooja Metal Enterprises, Qingyuan Jintian, Rama Paper Mills, Scrap Metal Processors, Sharif Metals, the Shredder Company, the Sigma Group, Sunrise Metal Recycling, Taizhou Xinglitong Metal Industry Co., Toxics Link, Tung Tai, Uni-All, Vans Chemistry, Waste Management Corporation, and Yong Chang Processing. (Note: Chinese sources are listed with the names that they use in an English-language setting.)

Vaman Acharya, Alan Alpert, Raymond Alpert, Adriano Assi, Tetsuro Azuma, Alan Bachrach, Terry Baumsten, Scott Bergmeyer, Stu Block, Rich Brady, Lynn Brown, Tim Bryan, Joe Chen, Chen Liwen, Terry Chen, David Chiao, Freddy Cohen, Frank Coleman, Matt Coz, Albert Cozzi, Frank Cozzi, Ron Dalton, Mike Diehl, Michel Dubois, Guy Dumato, Ben Eisbart, Howard Farber, Jake Farber, Belly Feng, Scott Frederick, Leonard Fritz, Randy Fritz, Su Fung Ow Yong, Scott Gibble, Shane Gilbertson, Josh Goldstein, Randy Goodman, Dick Gordon, Kal Gordon, Louis Gordon, Tim Heffernan, Yao Wei Heng, Frank Huang, Ivy Huang, Tony Huang, Robin Ingenthron, Alex Jiang, Julie Ketchum, Ian Kimmer, Homer Lai, Wing Lai, Dennis Leburg, Mark Leder, James Li, Raymond Li, Alan Liu, Josh Lohman, Denny Luma, Priti Mahesh, Scott McDaniel, Carl Michaud, Venkatesha Murthy, Brian Nachlis, Joalton Newell, Scott Newell, Terry Ng, Jack Noe, Anil Panchmatiya, Sunil Panchmatiya, Jim Puckett, Dennis Reno, Jr., Dennis Reno, Sr., Kurt Richardson, Russ Rinn, Harvey Rosen, Raymond Sarmento, Salam Sharif, Shi Tong Qu, Dave Simonds, Jim Skipsie, Tim Spiro,

Dave Stage, Pravinbhai Timbadia, Melissa Tsai, Frances Veys, David Wallace, Jean Wang, Wang Jinglian, Wang Jiwei, Wang Qun Ying, Kyle Wiens, Yao Yei, the invaluable Johnson Zeng, and the millions of scrap peddlers who labor up and down China's streets every day, inspiring me with their resourcefulness and hard work.

Finally, I am deeply grateful to the many scrap men and scrap women who asked that their contributions to my scrap education, and this book, remain anonymous. You know who you are.

Additional thanks are due to Katherine Brown, Rob Schmitz and Lenora Chu, Jim and Deb Fallows, Helga Fresen, Mitchell Gordon, Tobin Harshaw, Andrew and Cindy Hill, Mara Hvistendahl, Steve Kaplan, Zara Kessler, Benjamin Lorch, M. D. "Mush" Oberman, Rachel Pollack, Scott and Weiping Satterfield, and my *jiejie*, Wo Ye.

In the course of working on this book I was lucky to report and write in places where I have family and friends willing to host me. Those days (weeks, months . . .) are among the most enjoyable I had while working on this often daunting project, in large part because the people who hosted me are among those closest to me. In alphabetical order, they are: Amy Minter and Michael Bachrach, Bruce and Joanne Gruen, Erik Öberg and Jody Lyle, Steve and Leia Simon, John Tan and Michelle Ku, and all of the Zemans: Ed, Jane, Rachel, Matthew, Max and Betty.

Very special thanks to Mr. Tan for the loan of his office, and Ms. Ku, Lim Swee Wan, and the Good Women of JCMS ProRewards for their hospitality (and all of that *nasi lemak*).

Finally, to my beloved wife, Christine—thank you for your unwavering support, for your critical eyes and ears, and for telling people that you "never knew garbage until I knew Adam Minter." You're a blessing, a sweetener, my hidden box of Barley in a warehouse full of Birch/Cliff.

Index

A Note on the Author

Adam Minter grew up in a family of scrap dealers in Minneapolis. He is the Shanghai correspondent for *Bloomberg World View* and a regular contributor to *Scrap* and *Recycling International*. His writing has appeared in the *Atlantic*, *Foreign Policy*, the *Los Angeles Times*, *Sierra*, and other publications. He and his wife divide their time between Shanghai, Kuala Lumpur, and the western suburbs of Minneapolis. *Junkyard Planet* is his first book.